T0321438

Emerging Trends in Advanced Spectroscopy

RIVER PUBLISHERS SERIES IN OPTICS AND PHOTONICS

Series Editors:

FRANK CHANG
Inphi Corporation
USA

MANIJEH RAZEGHI
Northwestern University
USA

KEVIN WILLIAMS
Eindhoven University of Technology
The Netherlands

Indexing: All books published in this series are submitted to the Web of Science Book Citation Index (BkCI), to SCOPUS, to CrossRef and to Google Scholar for evaluation and indexing.

The "River Publishers Series in Optics and Photonics" is a series of comprehensive academic and professional books which focus on the theory and applications of optics, photonics and laser technology.

Books published in the series include research monographs, edited volumes, handbooks and textbooks. The books provide professionals, researchers, educators, and advanced students in the field with an invaluable insight into the latest research and developments.

Topics covered in the series include, but are by no means restricted to the following:

- Integrated optics and optoelectronics
- Applied laser technology
- Lasers optics
- Optical Sensors
- Optical spectroscopy
- Optoelectronics
- Biophotonics photonics
- Nano-photonics
- Microwave photonics
- Photonics materials

For a list of other books in this series, visit www.riverpublishers.com

Emerging Trends in Advanced Spectroscopy

Editors

Yang Weiman

Beijing University of Chemical Technology, China

Jibin K.P.

Mahatma Gandhi University, India

Praveen G.L.

Mahatma Gandhi University, India

Sabu Thomas

Mahatma Gandhi University, India

Nandakumar Kalarikkal

Mahatma Gandhi University, India

River Publishers

Published, sold and distributed by:
River Publishers
Alsbjergvej 10
9260 Gistrup
Denmark

River Publishers
Lange Geer 44
2611 PW Delft
The Netherlands

Tel.: +45369953197
www.riverpublishers.com

ISBN: 978-87-7022-082-8 (Hardback)
⠀⠀⠀⠀⠀978-87-7022-081-1 (Ebook)

Contents

MODULE 1

MODULE 3

Preface

The importance and potential of the spectroscopic tool are snowballing day by day in all fields of science particularly in the material characterisation. It is very indispensable to know how the diverse tools used worldwide to arrive at a deduction about a material. The scientists make use of a large assemblage of characterisation techniques to nurture data linking the recital of materials to physical as well as chemical structure. All type of materials requires various kinds of characteristics. The anticipated properties are achieved by the appropriate selection of starting molecules and the extreme control on the morphological and mechanical properties. The characterisation and resultant understanding require inputs from, FTIR, UV, XRD, Raman, NMR and XPS etc. Other practices include photoacoustic spectroscopy, neutron scattering, fluorescence and mass spectrometry. In drawing upon so many diverse and specialized fields the material characterisation is important and the idea should share and communicate with one another in society.

This book deals with the application of various spectroscopic techniques in a range of fields and context. The first chapter deals with the spectroscopic characteristics of Graphitic structures. We all know how important graphite and its derivatives in the scientific world. Each day a new horizon of graphite materials will develop. In preceding chapters, each author discusses their work and application of various spectroscopic tools in their work.

The aim of this book is to help scientists, academicians and students in all over the world and industry to make effective use of diverse spectroscopic tools in solving scientific problems.

We are especially indebted to every one of them for their support and endurance which made this book possible. The Editors would like to thank the management team of River Publisher for the permission to publish this book and their support throughout the process. We owe our greatest debt of gratitude to Mahatma Gandhi University and Beijing University of Chemical Technology for their help and support.

List of Contributors

Amarkumar Bhatt, *R&D Division, Gujarat State Fertilizers and Chemicals Limited, Fertilizer Nagar, Vadodara 391750, Gujarat, India; E-mail: amarbhatt@gsfcltd.com*

Ajay Vasudeo Rane, *Composite Research Group, Department of Mechanical Engineering, Durban University of Technology, Durban 4000, South Africa; E-mail: ranea061@gmail.com*

Arpita Das, *Department of Nutrition, Dr. Bhupendra Nath Dutta Smriti Mahavidyalaya, Burdwan – 713407, West Bengal, India; E-mail: arpitadas_0202@yahoo.com*

Bangari Daruka Prasad, *Department of Physics, BMS Institute of Technology and Management, VTU-Belagavi, Bengaluru 560 064, Karnataka, India; E-mail: darukap@bmsit.in*

Christopher Thresiamm Mathew, *Department of Physics, Mar Ivanios College, Thiruvananthapuram 695015, Kerala, India; E-mail: drmathewct@gmail.com*

Dhamodaran Dhanalakshmi, *Raman Research Institute, Sadashivanagar, Bangalore 560080, India; E-mail: dhana@rri.res.in*

Dibyendu Roy, *Raman Research Institute, Sadashivanagar, Bangalore 560080, India; E-mail: droy@rri.res.in*

Elias Chatzitheodoridis, *National Technical University of Athens (NTUA), Greece; E-mail: chatzitheodoridis@icloud.com*

Hanumanthappa NagBhushana, *CNR Rao Centre for Advanced Materials Research, Tumkur University, Tumkur 572 103, Karnataka, India; E-mail: bhushanvlc@gmail.com*

Hema Ramachandran, *Raman Research Institute, Sadashivanagar, Bangalore 560080, India; E-mail: hema@rri.res.in*

Jayachandran Santhakumari Lakshmi, *Department of Physics, Mar Ivanios College, Thiruvananthapuram 695015, Kerala, India; E-mail: lakshmibipin@gmail.com*

Jesiya Susan George, *International and Inter University Centre for Nanoscience and Nanotechnology, Mahatma Gandhi University, Kottayam, Kerala, India; E-mail: jesiyasusan1996@gmail.com*

Jijimon K. Thomas, *Department of Physics, Mar Ivanios College, Thiruvananthapuram 695015, Kerala, India; E-mail: jkthomasmail.emrl@gmail.com; jkthomasemrl@yahoo.com*

Keloth Paduvilan Jibin, *1. International and Inter University Centre for Nanoscience and Nanotechnology, Mahatma Gandhi University, Kottayam, Kerala, India; 2. School of Chemical Sciences, Mahatma Gandhi University, Kottayam, Kerala, India; E-mail: jibinkp999@gmail.com*

Khateef Riazunnisa, *Department of Biotechnology and Bioinformatics, Yogi Vemana University, Kadapa, Andhra Pradesh, India; E-mail: khateefriaz@gmail.com, krbtbi@yogivemanauniversity.ac.in*

Krishnan Thyagarajan, *Department of Physics, JNTUA College of Engineering, Pulivendula 516 390, Andra Pradesh, India*

Krishnan Kanny, *Composite Research Group, Department of Mechanical Engineering, Durban University of Technology, Durban 4000, South Africa; E-mail: kannyk@dut.ac.za*

Maram Vidya Vani, *Department of Biotechnology and Bioinformatics, Yogi Vemana University, Kadapa, Andhra Pradesh, India; E-mail: Vidya.mkutty173@gmail.com*

Maheswar Swar, *Raman Research Institute, Sadashivanagar, Bangalore 560080, India; E-mail: mswar@rri.res.in*

Monalisa Mishra, *Department of Life Science, NIT Rourkela, Rourkela, Odisha 769008, India; E-mail: mishramo@nitrkl.ac.in; monalisamishra2010@gmail.com*

Neehara Alackal, *Department of Chemistry Deva Matha College, Mahatma Gandhi University, Kottayam, Kuravilangad, Kottayam 686633, Kerala, India; E-mail: neeharaa2015@gmail.com*

Nirmit Kantilal Sanchapara, *R&D Division, Gujarat State Fertilizers and Chemicals Limited, Fertilizer Nagar, Vadodara 391750, Gujarat, India; E-mail: nksanchapara@gsfcltd.com*

N. L. Jayalekshmy, *Department of Physics, Mar Ivanios College, Thiruvananthapuram 695015, Kerala, India; E-mail: jayale82@yahoo.co.in*

Prajitha Velayudhan, *1. International and Inter University Centre for Nanoscience and Nanotechnology, Mahatma Gandhi University, Kottayam, Kerala, India; 2. School of Chemical Sciences, Mahatma Gandhi University, Kottayam, Kerala, India; E-mail: prajipravi.11@gmail.com*

Sandeep Kollam, *Department of Physics, Mar Ivanios College, Thiruvananthapuram 695015, Kerala, India; E-mail: sandeepkollam@gmail.com*

Suresh Puthiyaveetil Othayoth, *R&D Division, Gujarat State Fertilizers and Chemicals Limited, Fertilizer Nagar, Vadodara 391750, Gujarat, India; E-mail: posuresh@gsfcltd.com*

Syeda Anjum Mobeen, *Department of Biotechnology and Bioinformatics, Yogi Vemana University, Kadapa, Andhra Pradesh, India; E-mail: anjummobeen@gmail.com*

Soheb Husenmiyan Shekh, *R&D Division, Gujarat State Fertilizers and Chemicals Limited, Fertilizer Nagar, Vadodara 391750, Gujarat, India; E-mail: sohebshekh19@gmail.com*

Sandeep Jasvantrai Parikh, *R&D Division, Gujarat State Fertilizers and Chemicals Limited, Fertilizer Nagar, Vadodara 391750, Gujarat, India; E-mail: sjparikh@gsfcltd.com*

Sabu Thomas, *1. International and Inter University Centre for Nanoscience and Nanotechnology, Mahatma Gandhi University, Kottayam, Kerala, India; 2. School of Chemical Sciences, Mahatma Gandhi University, Kottayam, Kerala, India; E-mail: sabuthomas@mgu.ac.in*

Sam Solomon, *Department of Physics, Mar Ivanios College, Thiruvananthapuram 695015, Kerala, India; E-mail: sam.solomon@mic.ac.in; samdmrl@yahoo.com*

Sanjukta Roy, *Raman Research Institute, Sadashivanagar, Bangalore, 560080, India; E-mail: sanjukta@rri.res.in*

Saptarishi Chaudhuri, *Raman Research Institute, Sadashivanagar, Bangalore 560080, India; E-mail: srishic@rri.res.in*

Shilpa Bothra, *Department of Applied Chemistry, SV National Institute of Technology (SVNIT), Surat-395007, Gujarat, India; E-mail: bothrashilpa7@gmail.com*

Sohan Jheeta, *Network of Researchers on the Chemical Evolution of Life (NoR CEL), UK; E-mail: sohan@sohanjheeta.com*

Steffy Maria Jose, *Department of Physics, Mar Ivanios College, Thiruvananthapuram 695015, Kerala, India; E-mail: steffymaria09@gmail.com*

Suban K. Sahoo, *Department of Applied Chemistry, SV National Institute of Technology (SVNIT), Surat-395007, Gujarat, India; E-mail: suban_sahoo@rediffmail.com; sks@chem.svnit.ac.in*

Tandrima Chaudhuri, *Department of Chemistry, Dr. Bhupendranath Dutta Smriti Mahavidyalaya, Burdwan, West Bengal, India; E-mail: tanchem_bu@yahoo.co.in*

V. K. B. Kota, *Physical Research Laboratory, Ahmedabad 380009, India; E-mail: vkbkota@prl.res.in*

Yesoda Velukutty Swapna, *Department of Physics, Mar Ivanios College, Thiruvananthapuram 695015, Kerala, India; E-mail: swapnamct08@gmail.com*

List of Figures

List of Tables

List of Abbreviations

μm	micrometer
^1H NMR	Proton NMR
Ag^+	Silver ion
$AgNO_{3-}$	Silver nitrate
AgNPs	Silver nanoparticles
ATR	Attenuated total reflection
B1	0.5wt% carbon black filled Poly (lactic acid)
B2	1.0wt% carbon black filled Poly (lactic acid)
B3	1.5wt% carbon black filled Poly (lactic acid)
B4	2.0wt% carbon black filled Poly (lactic acid)
B5	2.5wt% carbon black filled Poly (lactic acid)
CB	Carbon Black
CH_3CN	Acetonitrile
CMS	Carboxy Methyl Cellulose
DBP	Dibutyl Phthalate
DTA	Differential Thermal Analysis
EDS	Energy Dispersive Spectroscopy
ESR	Equivalent Series Resistance
FIC	Fast Ion Conductor
FT	Fourier Transform
FTIR	Fourier Transform Infrared Spectrometer
FWHM	Full-Width-at-Half-Maximum
G	Conductance
g	gram
HOMO	Highest Occupied Molecular Orbital
HRTEM	High-Resolution Transmission Electron Microscopy
ICDD	International Centre for Diffraction Data
IDP	Interstellar Dust Particle
IR	Infrared
IRT	Infrared Transparent
IS	Impedance Spectroscopy

IS	Indian Standard
ISM	Interstellar Medium
JCPDS	Joint Committee on Powder Diffraction Standards
M	Molar
mM	milimolar
MB	Methylene Blue dye
MOSFET	Metal Oxide Semiconductor Field Effect Transistor
Neat PLA	Neat Poly (lactic acid)
NH_3–N	Ammonia – Nitrogen
nm	nanometer
NO_3^-–N	Nitrate – Nitrogen
NoR CEL	Network of Researchers on the Chemical Evolution of Life
NPs	Nanoparticles
NTUA	National Technical University of Athens, Greece
PDF	Powder Diffraction File
PL	Photoluminescence
PLA	Poly (lactic acid)
PO_4^3–P	Phosphate – Phosphorus
PSD	Particle Size Distribution
PVA	Poly Vinyl Alcohol
QPE	Constant Phase Element
SAED	Selected Area Electron Diffraction Pattern
SDBS	Sodium Dodecyl Benzene Sulfonate
SEM	Scanning Electron Microscopy
SLS	Sodium Lauryl Sulfonate
SOFCs	Solid Oxide Fuel Cells
TBAF	Tetrabutyl Ammonium Fluoride
TEM	Transmission Electron Microscope
TGA	Thermogravimetric Analysis
UV	Ultraviolet
UV-Vis	Ultraviolet-Visible
XRD	X-ray Diffraction
Y_2O_3	Yttrium Oxide (Yttria)
ZnO	Zinc Oxide

MODULE 1

1

Spectroscopic Characterizations of Graphitic Structures

Jesiya Susan George[1], Keloth Paduvilan Jibin[1,2],
Prajitha Velayudhan[1,2] and Sabu Thomas[1,2,*]

[1]International and Inter University Centre for Nanoscience and
Nanotechnology, Mahatma Gandhi University, Kottayam, Kerala, India
[2]School of Chemical Sciences, Mahatma Gandhi University, Kottayam,
Kerala, India
E-mail: sabuthomas@mgu.ac.in
*Corresponding Author

Graphite and its derivatives are the most intentional materials today. Specifically, the two-dimensional graphene has tremendous responsiveness because of its exceptional properties that include adsorption of gas molecules, high carrier mobility, Hall effect at room temperature, infrequent magnetic properties, ballistic conduction of charge carriers, highly elastic nature, and remarkable surface area. Therefore, characterization of graphene is an important part of graphene exploration which comprises different microscopic and spectroscopic methods. In characterization, the main motive is to determine the number of layers and the purity of sample by evaluating the absence or presence of defects. We discuss some of the spectroscopic aspects in this chapter.

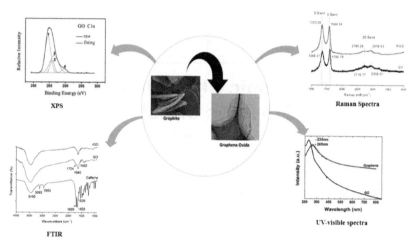

Graphical Abstract

1.1 Introduction

1.1.1 Graphite

Graphite is a naturally occurring stable form of carbon, in which each carbon atom is arranged in a hexagonal structure with SP^2 hybridization. Graphite has a layered planar structure. The single layer of graphite is called graphene, and each layer is held together by covalent bonds and Vander wall forces. Alpha/hexagonal graphite and beta/rhombohedral graphite are the two types of graphite, and both the alpha and beta forms of graphite show similar physical properties; however, the difference is only in layer stacking and both types are inter-convertable. Alpha form can be converted to beta form by mechanical treatment, whereas beta form can be converted to alpha form by thermal treatment above $1300°C$ [1–4]. Figure 1.1 shows the interconversion of α graphite to β graphite and vice versa.

Graphite exhibits higher thermal stability, electrical conductivity, thermal conductivity, and anisotropic nature. Because of the combination of these excellent properties, graphite and graphite-based composites are widely used in batteries, fuel cells, brake lining, and lubricants [5–6].

1.1.2 Graphene Oxide and Graphene

Graphene and graphene oxide are the two known derivatives of natural graphite. The oxidized form of graphite is known as graphene oxide, whereas

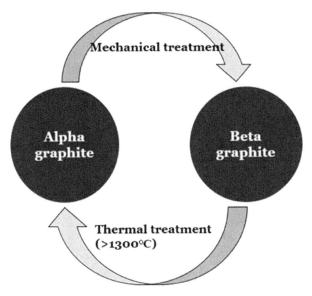

Figure 1.1 Interconversion of α-graphite to β-graphite and vice versa.

the one-atom-thick layer of graphite is called graphene, the former was discovered in 1859 by the chemist "Benjamin Collins Brodie" and the latter was discovered in 2004 by "Geim and Novoselov" [5].

Graphene oxide (GO) can be considered as the heavily oxygenated form of graphite, obtained by the chemical oxidation of graphite using various oxidizing agents such as $KMnO_4$, $KClO_3$, and K_2FeMO_4. During the last decade, several methods are used for the preparation of graphene oxide (Brodie method, Staudenmaier Hummer method, and Tour method). Brodie method was reported in 1859, which involves the oxidation of graphite using potassium perchloride ($KClO_3$) in fuming nitric acid (HNO_3). But in 1898, Staudenmaier made some modifications in the Brodie method. He increased the acidity of the reaction mixture by incorporating concentrated sulfuric acid (H_2SO_4). Brodie and Staudenmaier method yields more oxidation; however, the formation of toxic ClO_2 is the limitation of both methods. In 1958, Hummers and Offeman proposed a new method for graphite oxide synthesis, which also achieves the same extent of oxidation of Brodie method and Staudenmaier method. This method involves the oxidation of graphite using potassium permanganate and sodium nitrate in con. sulfuric acid; later, the reaction was terminated using hydrogen peroxide and washed several times with distilled water and basic solution until it becomes neutral. In 2010, a group of researchers from Rice University improved the Hummers' method by replacing sodium nitrite in a mixture of phosphoric acid and sulfuric acid (1:9) and increasing the amount of $KMnO_4$, called Tour method or improved Hummers method. This method yields graphene oxide with higher degree of oxidation and hydrophilicity. In addition to these three methods, graphene oxide can also be prepared by other methods such as [5–9]:

- Sun method [10]
- Peg method [11]
- Free water oxidation method
- Four-step method [12]

Graphene is the one-atom-thick layer of graphite, consisting of SP^2 hybridized carbon atoms. It was discovered in 2014 by Andre Geim and Kostya Novoselov and gained the interest of researchers because of its unusual mechanical, electrical, optical, and other properties. Reduction of graphene oxide, chemical vapor deposition, and mechanical exfoliation of graphite are the most popular techniques used for the preparation of graphene [13–16]. Figures 1.2 and 1.3 illustrate about the preparation methods and potential applications of graphitic derivatives respectively.

Graphene and graphene oxide has lot of potential applications in almost every field such as water purification [17], energy storage [18], super capacitors [19], drug delivery [20], and sensors [21].

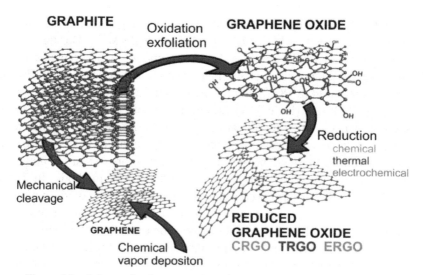

Figure 1.2 Schemes for the preparation of graphene and GO [14] open access.

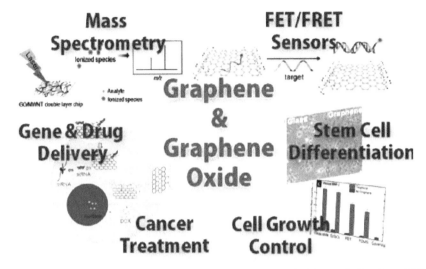

Figure 1.3 Applications of GO and graphene [40]. Reproduced with permission from ACS.

Property	Graphite	Graphene Oxide	Graphene
Synthesis method	Obtained naturally	Oxidation of graphite via Hummers' method, Tour method, etc.	Chemical vapor deposition, exfoliation of graphite, thermal decomposition of SiC
Thermal conductivity (W/cm-K)	300.06		$2000–4000 \text{ Wcm}^{-1}\text{K}^{-1}$
Electron mobility $(\text{Cm}^2\text{V}^{-1}\text{s}^{-1})$	20×10^3	Insulator	10000–50000
Wettability	Hydrophobic	Hydrophilic	Hydrophobic
Hybridization	SP^2	SP^3	SP^2
Carbon to oxygen ratio	No oxygen	2–4	No oxygen
Structure	Multilayers	Sheet like	Sheet like

1.2 Spectral Characterizations of Graphite and Graphite Derivatives

1.2.1 UV-Visible Spectroscopy

UV-visible spectroscopy is a robust characterization technique that can be used for the structure elucidation of organic molecules and also for both

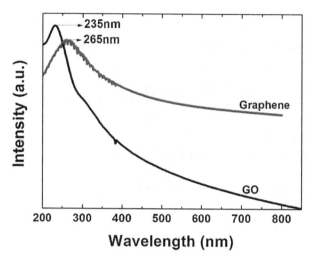

Figure 1.4 UV-visible absorption spectra of graphene and graphene oxide [39]. Reproduced with permission from Elsevier.

qualitative and quantitative analyses. It is also useful for the characterization of graphite and its derivatives. Fatima et al. discussed the facile hydrothermal reduction of graphene oxide. They showed the UV-Vis absorption spectra of neat graphene oxide and reduced graphene oxide.

Figure 1.4 is a typical UV spectrum of Graphene and Graphene oxide. Aqueous dispersion of graphene oxide shows a characteristic absorption peak at 235 nm, but after reaction the peak shifts to 272 nm, which corresponds to $\pi-\pi^*$ transitions of the remaining SP^2 C=C bonds and indicates the reduction of graphene oxide, thereby restoring the electronic conjugation within the sheets of reduced graphene oxide [21, 22, 39].

Generally, in the UV-visible absorption spectra of graphene oxide, peak at 232 nm corresponds to $\pi-\pi^*$ transition of aromatic C=C bonds, whereas $n-\pi^*$ transition of CHO bonds gives absorption at 298 nm. After the reduction, it will shift into red region. That is, the 232 nm shifts to 269 nm and the other peak is completely removed [23].

1.2.2 Raman Spectroscopy

Raman spectroscopy is used to understand the structural hybridization changes in graphite and its derivatives. Graphite consists of SP^2 hybridized carbon atoms, but on conversion to graphene oxide and graphene, the hybridization changes to SP^3 and SP^2, respectively.

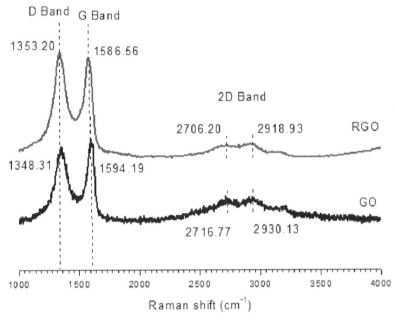

Figure 1.5 Raman spectra of GO and RGO [26]. Reproduced with permission from AIP Publishing.

Hidayah et al. reported the preparation and characterization of graphene oxide and reduced graphene via improved Hummers' method and chemical reduction using hydrazine hydrate. Figure 1.5 shows the Raman spectra of graphene oxide and reduced graphene oxide. Raman spectra of graphitic structures are generally distinguished by two main peaks/bands: G band and D band. G band is formed by the first-order scattering of E_{2g} phonons by SP^2 carbon atoms, whereas D bands are formed by the breathing mode of j-point photons of A_{1g} symmetry.

Generally, G bands appear at 1575 cm^{-1} and D bands appear at 1350 cm^{-1}. In Figure 1.5, the G band and the D band of graphene oxide were observed at 1549 cm^{-1} and 1348 cm^{-1}, respectively. But after reduction using hydrazine hydrate, the G band is shifted to 1353 cm^{-1} and the D band appears at 1386 cm^{-1}. In the Raman spectra of graphitic structures, two-dimensional (2D) bands are used to determine the layers. Here the 2D band of both graphene oxide and reduced graphene oxide appears in the range of 2700 cm^{-1} and is multilayered in nature, since the monolayer graphene was normally observed at 2679 cm^{-1} in the spectra. The 2D shifting in graphene oxide is due to the presence of oxygen moieties that prevent the single sheets to stack but RGO shows this 2D band at 2706.20 cm^{-1}; this is because the RGO starts to stack after reduction.

Another important parameter obtained from Raman spectra is the I_D/I_G ratio. Oxidation of graphite to graphene oxide is accomplished by the addition of oxygen moieties and hybridization change. Graphite has SP^2 hybridization, but after oxidation it converts to SP^3; however, the reduction of graphene oxide again changes the hybridization of carbon to SP^3; therefore, the reduction can be considered as a re-graphitization step, and that is why graphite and graphene show similar characteristics. The I_D/I_G ratio of graphene oxide is 0.84; the reduced graphene oxide has an I_D/I_G ratio almost similar to graphite [24–34].

However, Madhusmita et al. reported the Raman studies of chemically and thermally reduced graphene oxide. The prepared graphene oxide was chemically reduced via hydrazine and thermally reduced at 900°C, 1000°C, and 1100°C for about 1 minute. Figure 1.6 shows the Raman spectrum of thermally exfoliated graphite. As we already discussed that I_D/I_G ratio is

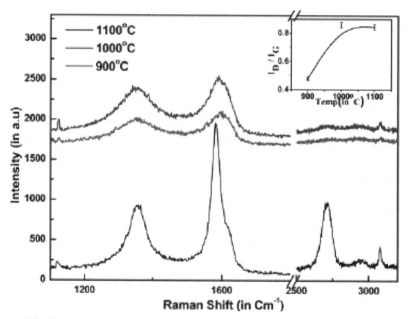

Figure 1.6 Raman spectra of thermally reduced graphene oxide (TRGO) [38]. Reproduced with permission from AIP Publishing.

a parameter obtained from Raman spectra, in Figure 1.5 the I_D/I_G ratio increases as the exfoliation temperature increases from $900°C$ to $1100°C$. Also the sample exfoliated at $900°C$ shows a well-defined 2D band at about $2600–2700$ cm^{-1} but as the exfoliation temperature increases the 2D band becomes broad and starts to vanish [38].

1.2.3 X-Ray Photoelectron Spectroscopy

X-ray photoelectron spectroscopy (XPS) or electron spectroscopy for chemical analysis (ESCA) is a powerful surface sensitive quantitative technique, widely used for the elemental composition analysis of graphitic materials. It can also be used for the determination of empirical formula, chemical state, and electronic state of the elements within a material. Reduced graphene oxide and graphene oxides consist of SP^2 and SP^3 carbon atoms, respectively. Compounds with high concentration of SP^2 carbon give a C1s spectrum with broad, asymmetric tail towards higher binding energy, whereas compounds with high concentration of SP^3 carbon atoms provide C1s spectrum with more symmetric shape and will also be shifted to higher binding energy.

Feng et al. employed XPS for the elemental analysis of graphene oxide and graphene. Graphene oxide is the massively oxygenated form of graphite with oxygen functionalities such as hydroxyl, epoxy, and carbonyl moieties. Figure 1.7(a) represents the XPS spectra of graphene oxide, which exhibits four different peaks centered at $284.0–290.0$ eV. The peak at 284.5 eV in the XPS spectra corresponds to the presence of C$=$/C$-$C in aromatic rings, whereas 286.4, 287.8, and 289.0 eV peaks correspond to the presence of epoxy carbonyl and carboxyl groups in the graphene oxide.

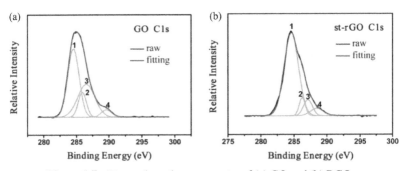

Figure 1.7 X-ray photoelectron spectra of (a) GO and (b) RGO.

Figure 1.7(b) shows the XPS spectra of graphene oxide after reduction using starch. The intensities of all C1s peaks of carbon binding to oxygen, particularly –C–O– peaks (epoxy, carbonyl), decrease drastically after reduction, which explains that the oxygen-containing moieties are reduced after reduction [35].

1.2.4 Fourier Transform Infrared Spectroscopy

Fourier transform infrared spectroscopy (FTIR) is a non-destructive characterization technique used to identify the functional groups in it. Chemical bonds vibrate at characteristic frequencies, when they are exposed to infrared radiation (400–4000 cm^{-1}), and they absorb the radiation at frequencies that match their vibration modes.

Thu et al. employed caffeine for the reduction of graphene oxide, an alkaloid mainly extracted from the leaves and seeds of coffee plant. Caffeine is chemically trimethylxanthine (1,3,7-trimethylxanthine); it can be used as a green reductant for graphene oxide, as it is biocompatible, nontoxic, and ease of availability at industrial level.

The characteristic peak of caffeine observed at 3093 and 2953 cm^{-1} represents N–H and C–H stretching, respectively. It also shows two strong bands that correspond to carbonyl groups at 1699 and 1659 cm^{-1}. Also, C=N shows absorption band at 1539 cm^{-1}. A typical FTIR spectrum of graphene oxide is mainly divided into three parts, an intense and very broad absorption band as shown in Figure 1.8 in the 3600–2400 cm^{-1} region, most recognizable absorption bands at 1723 cm^{-1} and 1619 cm^{-1} in the middle of the spectrum, and a bunch of overlapping signals in the fingerprint region; 3600–2400 cm^{-1} is due to OH stretching; and the pair of absorption bands in the middle of the spectrum at 1723 cm^{-1} and 1619 cm^{-1} is the signature of the FTIR spectra of GO. The position of the first peak sometimes varies between 1719 cm^{-1} and 1734 cm^{-1}. The 1723 cm^{-1} band is due to the stretching mode of carbonyl group present in graphene oxide [36, 37].

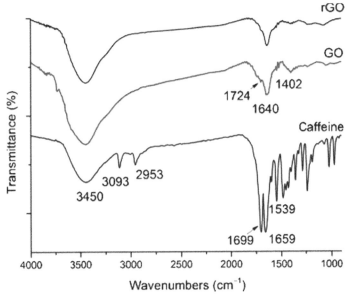

Figure 1.8 FTIR of GO, RGO, and caffeine [36]. Reproduced with permission from Springer.

1.2.5 X-Ray Diffraction

Manchala et al. introduced a novel and efficient method of green reduction of graphene oxide using eucalyptus bark extract. The powder XRD of graphite and its derivatives is carried out to determine the interlayer spacing and atomic structures. Figure 1.9 represents the results of the X-ray diffraction studies of graphite, GO, and graphene. Natural graphite shows a strong and intensive signature peak at 26.6° which is due to the interlayer (d) spacing of 0.335 nm. But after oxidization, the diffraction peak shifts to lower angle 12.2° (d$_{spacing}$ 0.725), and it seems to be more broad. The supporting information for the efficient reduction of GO by green extract was obtained from the XRD results of E-graphene, the peak at 12.2° of GO disappeared after reduction, and a new broad peak at 25° was obtained with a d-spacing of 0.356 nm.

The d-spacing of graphene oxide is relatively high, which is almost twice of the natural graphite, which is a direct result of the oxidation and intercalation of natural graphite. However, the d-spacing of graphene is appreciably decreased which point outs the removal of oxygen functionalities [41, 42].

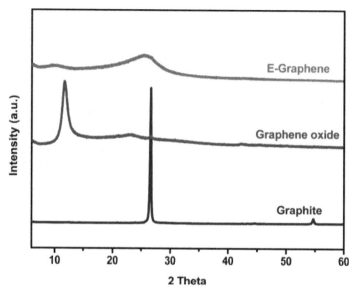

Figure 1.9 XRD of graphite, graphene oxide, and graphene [42]. Reproduced with permission from ACS.

1.3 Conclusion

In this review chapter, we discussed extensively the synthesis and spectral characterization of graphite and its structural derivatives based on the fundamental understanding of chemistry and material science. The synthesis and characterization of graphitic structures are very important because of the exceptionally high demand in various fields such as electronic devices, energy technology, water purification, dielectric materials, photo catalysis, solar cells, biomedical field, and tyre engineering. Also the UV, Raman, XPS FTIR X-ray, and diffraction studies were discussed.

Acknowledgement

First author acknowledge the financial support from DRDO-NRB (NRB-429/MAT/18-19), second and third authors greatly acknowledge the financial support from Department of Science and Technology through Nanomission Project (SR/NM/NT-1054/2015(G)).

References

[1] Delhaes, P. (2014). *Graphite and precursors*. CRC Press.

[2] Lipson, H. and Stokes, A. R. (1942). A new structure of carbon. *Nature*, 149(3777), 328.

[3] Wyckoff, R. W. G. (1964). *Crystal structures*. Krieger.

[4] Gold, V. I. C. T. O. R., Loening, K. L., McNaught, A. D. and Sehmi, P. (1987). *International Union of Pure and Applied Chemistry Compendium of Chemical Terminology IUPAC Recommendations*. Blackwell Scientific Publications Limited.

[5] Dreyer, D. R., Park, S., Bielawski, C. W. and Ruoff, R. S. (2010). The chemistry of graphene oxide. *Chemical Society Reviews*, 39(1), 228–240.

[6] Dimiev, A. M. and Tour, J. M. (2014). Mechanism of graphene oxide formation. *ACS Nano*, 8(3), 3060–3068.

[7] Hummers Jr, W. S. and Offeman, R. E. (1958). Preparation of graphitic oxide. *Journal of the American Chemical Society*, 80(6), 1339.

[8] Staudenmaier, L. (1898). Method for the preparation of graphitic acid. *Ber Dtsch Chem Ges*, 31, 1481–1487.

[9] Marcano, D. C., Kosynkin, D. V., Berlin, J. M., Sinitskii, A., Sun, Z., Slesarev, A. and Tour, J. M. (2010). Improved synthesis of graphene oxide. *ACS Nano*, 4(8), 4806–4814.

[10] Sun, L. and Fugetsu, B. (2013). Mass production of graphene oxide from expanded graphite. *Materials Letters*, 109, 207–210.

[11] Peng, L., Xu, Z., Liu, Z., Wei, Y., Sun, H., Li, Z. and Gao, C. (2015). An iron-based green approach to 1-H production of single-layer graphene oxide. *Nature Communications*, 6, 5716.

[12] Geim, A. K. and Novoselov, K. S. (2010). The rise of graphene. *Nanoscience and Technology: A Collection of Reviews from Nature Journals*, pp. 11–19.

[13] Shareena, T. P. D., McShan, D., Dasmahapatra, A. K. and Tchounwou, P. B. (2018). A review on graphene-based nanomaterials in biomedical applications and risks in environment and health. *Nano-Micro Letters*, 10(3), 53.

[14] Rowley-Neale, S. J., Randviir, E. P., Dena, A. S. A. and Banks, C. E. (2018). An overview of recent applications of reduced graphene oxide as a basis of electroanalytical sensing platforms. *Applied Materials Today*, 10, 218–226.

[15] Velicky, M., Bradley, D. F., Cooper, A. J., Hill, E. W., Kinloch, I. A., Mishchenko, A. and Worrall, S. D. (2014). Electron transfer kinetics on mono-and multilayer graphene. *ACS Nano*, 8(10), 10089–10100.

[16] Yuan, W., Zhou, Y., Li, Y., Li, C., Peng, H., Zhang, J. and Shi, G. (2013). The edge-and basal-plane-specific electrochemistry of a single-layer graphene sheet. *Scientific Reports*, 3, 2248.

[17] Han, Y., Xu, Z. and Gao, C. (2013). Ultrathin graphene nanofiltration membrane for water purification. *Advanced Functional Materials*, 23(29), 3693–3700.

[18] Pumera, M. (2011). Graphene-based nanomaterials for energy storage. *Energy and Environmental Science*, 4(3), 668–674.

[19] Liao, Q., Li, N., Jin, S., Yang, G. and Wang, C. (2015). All-solid-state symmetric supercapacitor based on Co_3O_4 nanoparticles on vertically aligned graphene. *ACS Nano*, 9(5), 5310–5317.

[20] Goenka, S., Sant, V. and Sant, S. (2014). Graphene-based nanomaterials for drug delivery and tissue engineering. *Journal of Controlled Release*, 173, 75–88.

[21] Jiang, H. (2011). Chemical preparation of graphene−based nanomaterials and their applications in chemical and biological sensors. *Small*, 7(17), 2413–2427.

[22] Mahata, S., Sahu, A., Shukla, P., Rai, A., Singh, M. and Rai, V. K. (2018). The novel and efficient reduction of graphene oxide using *Ocimum sanctum* L. leaf extract as an alternative renewable bioresource. *New Journal of Chemistry*, 42(24), 19945–19952.

[23] Si, Y. and Samulski, E. T. (2008). Exfoliated graphene separated by platinum nanoparticles. *Chemistry of Materials*, 20(21), 6792–6797.

[24] Thakur, S. and Karak, N. (2012). Green reduction of graphene oxide by aqueous phytoextracts. *Carbon*, 50(14), 5331–5339.

[25] Dresselhaus, M. S., Jorio, A., Hofmann, M., Dresselhaus, G. and Saito, R. (2010). Perspectives on carbon nanotubes and graphene Raman spectroscopy. *Nano Letters*, 10(3), 751–758.

[26] Hidayah, N. M. S., Liu, W. W., Lai, C. W., Noriman, N. Z., Khe, C. S., Hashim, U. and Lee, H. C. (2017). Comparison on graphite, graphene oxide and reduced graphene oxide: Synthesis and characterization. In *AIP Conference Proceedings* (Vol. 1892, No. 1, p. 150002). AIP Publishing.

[27] Tuinstra, F. and Koenig, J. L. (1970). Raman spectrum of graphite. *The Journal of Chemical Physics*, 53(3), 1126–1130.

[28] Low, F. W., Lai, C. W. and Hamid, S. B. A. (2015). Easy preparation of ultrathin reduced graphene oxide sheets at a high stirring speed. *Ceramics International*, 41(4), 5798–5806.

[29] Stankovich, S., Piner, R. D., Nguyen, S. T. and Ruoff, R. S. (2006). Synthesis and exfoliation of isocyanate-treated graphene oxide nanoplatelets. *Carbon*, 44(15), 3342–3347.

[30] Wang, Y., Shi, Z. and Yin, J. (2011). Facile synthesis of soluble graphene via a green reduction of graphene oxide in tea solution and its biocomposites. *ACS Applied Materials and Interfaces*, 3(4), 1127–1133.

[31] Guo, S., Wen, D., Zhai, Y., Dong, S. and Wang, E. (2011). Ionic liquid–graphene hybrid nanosheets as an enhanced material for electrochemical determination of trinitrotoluene. *Biosensors and Bioelectronics*, 26(8), 3475–3481.

[32] Kudin, K. N., Ozbas, B., Schniepp, H. C., Prud'Homme, R. K., Aksay, I. A. and Car, R. (2008). Raman spectra of graphite oxide and functionalized graphene sheets. *Nano Letters*, 8(1), 36–41.

[33] Casiraghi, C., Pisana, S., Novoselov, K. S., Geim, A. K. and Ferrari, A. C. (2007). Raman fingerprint of charged impurities in graphene. *Applied Physics Letters*, 91(23), 233108.

[34] Malard, L. M., Pimenta, M. A. A., Dresselhaus, G. and Dresselhaus, M. S. (2009). Raman spectroscopy in graphene. *Physics Reports*, 473(5–6), 51–87.

[35] Feng, Y., Feng, N. and Du, G. (2013). A green reduction of graphene oxide via starch-based materials. *RSC Advances*, 3(44), 21466–21474.

[36] Vu, T. H. T., Tran, T. T. T., Le, H. N. T., Nguyen, P. H. T., Bui, N. Q. and Essayem, N. (2015). A new green approach for the reduction of graphene oxide nanosheets using caffeine. *Bulletin of Materials Science*, 38(3), 667–671.

[37] Garrigues, J. M., Bouhsain, Z., Garrigues, S. and de la Guardia, M. (2000). Fourier transform infrared determination of caffeine in roasted coffee samples. *Fresenius' Journal of Analytical Chemistry*, 366(3), 319–322.

[38] Sahoo, M., Antony, R. P., Mathews, T., Dash, S. and Tyagi, A. K. (2013). Raman studies of chemically and thermally reduced graphene oxide. In AIP Conference Proceedings (Vol. 1512, No. 1, pp. 1262–1263). AIP Publishing.

[39] Johra, F. T., Lee, J. W. and Jung, W. G. (2014). Facile and safe graphene preparation on solution based platform. *Journal of Industrial and Engineering Chemistry*, 20(5), 2883–2887.

[40] Chung, C., Kim, Y. K., Shin, D., Ryoo, S. R., Hong, B. H. and Min, D. H. (2013). Biomedical applications of graphene and graphene oxide. *Accounts of Chemical Research*, 46(10), 2211–2224.

[41] Bo, Z., Shuai, X., Mao, S., Yang, H., Qian, J., Chen, J. and Cen, K. (2014). Green preparation of reduced graphene oxide for sensing and energy storage applications. *Scientific Reports*, 4, 4684.

[42] Manchala, S., Tandava, V. S. R. K., Jampaiah, D., Bhargava, S. K. and Shanker, V. (2019). Novel and highly efficient strategy for the green synthesis of soluble graphene by aqueous polyphenol extracts of eucalyptus bark and its applications in high-performance supercapacitors. *ACS Sustainable Chemistry and Engineering*, 7(13), 11612–11620.

2

Transport Properties of Biotemplate-Mediated Nano Mixed Ferrites and Their Applications to Nanoelectronics

Bangari Daruka Prasad[1,*], Hanumanthappa NagBhushana[2,*] and Krishnan Thyagarajan[3]

[1]Department of Physics, BMS Institute of Technology and Management, VTU-Belagavi, Banglore 560 064, Karnataka, India
[2]CNR Rao Centre for Advanced Materials Research, Tumkur University, Tumkur 572 103, Karnataka, India
[3]Department of Physics, JNTUA College of Engineering, Pulivendula 516 390, Andhra Pradesh, India
E-mail: darukap@bmsit.in; bhushanvlc@gmail.com
*Corresponding Authors

Bio-mediated nanosized nickel-doped zinc ferrites were prepared by low temperature solution combustion method. The effect of the incorporation of Ni^{2+} ions on the structural, morphological, and transport properties was investigated. XRD data confirm the cubic spinel structure with lattice constant of 8.412Å. The refined parameters were of good agreement with the earlier reports. The particles were almost spherical and agglomerated which were confirmed using SEM. The average crystallite size was found to be $\sim$27 nm and this was confirmed using the results of both XRD and TEM. Infrared spectroscopy study confirmed the presence of strong metal oxide bonding. The frequency-dependent dielectric constant is found to be of the order of 10^4 at low frequency and almost 1 at higher frequency range. These studies suggest non-Debye type of dielectric relaxation in these samples. The prepared samples were quite useful for nanoelectronics applications.

2.1 Introduction

Nanosized magnetic particles were presently used in recording technology and biomedical applications because of their interesting structural and magnetic properties [1, 2]. It has been reported that the incorporation of metal ions in the spinel ferrite lattice strongly affects the structural, electronic, and magnetic properties of spinel ferrites. The substitution may prefer tetrahedral A-site or octahedral B-site [3]. Hence, the cation distribution changes the materials behavior. Among the various synthesis techniques, *viz* co-precipitation, microemulsion, ceramic method, hydrothermal, etc., it is important to know that the bio-mediated solution combustion method has many advantages such as low processing cost, energy efficiency, and high production rate. Further, the fuel/surfactant used in this method is the natural extracts. This chapter deals with the preparation of nickel-doped zinc ferrite using *Mimosa pudica* leaf extract–mediated solution combustion, characterization, and dielectric studies of the prepared samples.

2.2 Experimental Section

Fresh leaves of *Mimosa pudica* were washed thoroughly, dried, and kept under shade for 15–20 days at room temperature. The fresh solution of the plant extract is prepared using successive solvent extraction made in Soxhlet apparatus and refluxed with deionized water. The leaf extract has been used as a fuel/surfactant along with high pure zinc, nickel, and iron nitrates. The stoichiometric quantities of the precursors were stirred thoroughly along with deionized water and kept in preheated Muffle furnace. After smoldering type of burning process, the end product was crushed and heat treated at 700°C for 2 h. The characterization was carried out using advanced characterization techniques like XRD, FTIR, SEM, and TEM. For dielectric measurements, the pellets of known dimensions were prepared, and the surfaces were silvered for better Ohmic contacts. The dielectric measurements were carried out using Wayne Kerr 6500 B precession impedance analyzer in the frequency range 1–10 MHz.

2.3 Results and Discussion

Figure 2.1 shows the X-ray diffraction pattern (XRD) of $Zn_{1-x}Ni_xFe_2O_4$ ($x = 0.1, 0.3, 0.5, 0.7$, and 0.9) which confirms the cubic structure with space group Fd-3m (227) of the sample with JCPDS No. 52-0277. The measured

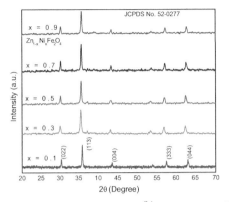

Figure 2.1 XRD patterns of Ni^{2+}-doped $ZnFe_2O_4$.

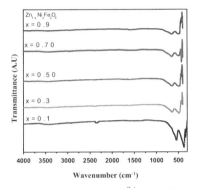

Figure 2.2 FTIR spectra of Ni^{2+}-doped $ZnFe_2O_4$.

average crystallite size from Scherrer formula is $\sim$30 nm. Figure 2.2 shows the spectra of $Zn_{1-x}Ni_xFe_2O_4$, and the transmittance peak at 500–600 cm^{-1} corresponds to intrinsic stretching vibrations of the metal at the tetrahedral site, $M_{tetra}\leftrightarrow O$, whereas the octahedral-metal stretching $M_{octa}\leftrightarrow O$ is at 400–385 cm^{-1}. The increase in the lattice constant affects the $Fe^{3+}-O^{2-}$ stretching vibrations and is a major cause for change in band positions. Figure 2.3a shows the agglomerated, fluffy, and porous nature of the sample using SEM image. Figure 2.3b shows the agglomerated TEM image with the crystallite size of about 27 nm.

But the plant extract used helps in obtaining the superstructures of the samples due to the mesh of polysaccharides and protein chains of the extract. These contents trap the metal ions and help in bonding to provide the complex

Figure 2.3   (a) SEM image, (b) TEM image, and (c) SAED pattern of $Zn_{0.5}Ni_{0.5}Fe_2O_4$ nanocompound.

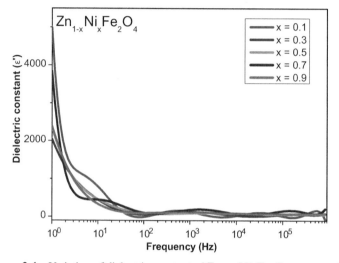

Figure 2.4   Variation of dielectric constant of $Zn_{1-x}Ni_xFe_2O_4$ nanopowder.

structures as shown in the SEM image. It can be explained on the basis of egg box model [4].

Figure 2.4 shows the variation of the dielectric constant with Ni^{2+} ions concentration at room temperature. The ϵ value decreases with an increase in Ni^{2+} ions concentration. This is because of increased space charge polarization and Koop's two-layered model [5]. The electrical conduction in ferrites can be explained in terms of the hopping of electrons between Fe^{2+} and Fe^{3+} ions at B-sites. The electrons reach the grain boundary by hopping and pile up due to their higher resistivity. This produces the space charge polarization. In this study, the Ni^{2+} ions concentration produces a change in the polarization due to cationic distribution. Increase in Ni^{2+} concentration pushes the Fe^{3+} ions into Fe^{2+} ions at B-site. The decrease in Fe^{3+} ion at the B-site decreases the charge mobility carrier.

2.4 Conclusions

The $Zn_{1-x}Ni_xFe_2O_4$ nanopowders were prepared by bio-mediated solution combustion method using *Mimosa pudica* L. extract. The XRD confirms the formation of the compound without any impurities and confirms the cubic crystallite phase. FTIR confirms the purity of the sample with strong metal–oxygen bonding within 600 cm^{-1} wave number. SEM and TEM images confirm the nano-regime and the agglomeration of the samples. The dielectric constant of these samples varies with Ni^{2+} content, and it was found to decrease with an increase in the Ni^{2+} content due to shifting of Fe^{3+} ions from tetrahedral to octahedral sites.

Acknowledgements

Author DP thanks the management and the Principal of BMSIT for their constant support and encouragement. Authors HN and BRB thank the DST-SERB (FAST TRACK) SR/FTP/PS – 135/2010, New Delhi, India, for providing financial assistance.

References

[1] C. Yao, Q. Zeng, G. F. Goya, T. Torres, J. Liu, H. Wu, M. Ge, J. Z. Jiang (2007). *J. Phys. Chem.* 111, 12274–12278.

[2] M. W. Mukhtar, M. Irfan, I. Ahmad, I. Ali, M. N. Akhtar, M. A. Khan, G. Abbas, M. U. Rana, A. Ali, M. Ahmad (2015). *J. Magn. Magn. Mater.* 381, 173–178.

[3] M. N. Akhtar, N. Yahya, A. Sattar, M. Ahmad, M. Idrees, M. H. Asif, M. A. Khan (2014). *Int. J. Appl. Ceram. Technol.* 1, 1–13.

[4] T. R. Lakshmeesha, M. K. Sateesh, B. Daruka Prasad, S. C. Sharma, D. Kavyashree, M. Chandrasekhar, H. Nagabhushana (2014). *Cryst. Growth Des.* 14, 4068–4079.

[5] A. I. Borhan, A. R. Iordan, M. N. Palamaru (2013). *Mater. Res. Bull.* 48, 2549–2556.

3

Structural, Optical and Impedance Spectroscopic Characterizations of $La_2Ti_2O_7$ Nanoceramic

Sandeep Kollam, Jijimon K. Thomas and Sam Solomon[*]

Department of Physics, Mar Ivanios College, Thiruvananthapuram 695 015, Kerala, India
E-mail: sam.solomon@mic.ac.in
[*]Corresponding Author

$La_2Ti_2O_7$ nanopowder is prepared through the single step autoignited combustion technique. The structural characterization is performed by X-ray diffraction (XRD) and vibrational spectroscopy. The optical studies are carried out using ultraviolet-visible (UV-Vis) and photoluminescence (PL) spectroscopic techniques. The XRD pattern shows that the structure of the compound is orthorhombic with a crystallite size of approximately 31 nm. Tauc's plot and the optical absorption spectrum are employed to calculate the optical band gap energy. The PL studies show that the emission belongs to the violet and red spectral area. Dielectric and electrical property measurements in the temperature range 500–800°C are performed. Complex impedance analysis is used to determine the grain and grain boundary effects on the dielectric behavior of the nanoceramic.

3.1 Introduction

The compounds with general formula $A_2B_2O_7$, have ferroelectric, multi-ferroic, and photocatalytic applications. They have pyrochlore- and perovskite-layered structures. The structure of perovskite compounds is characterized by perovskite slabs stacked along the 'a' axis, which are made up of corner sharing BO_6 octahedra and 12 coordinated A cations. Each slab is four octahedra thick and is linked to a neighboring slab by A cations lying near the boundary [1].

Herrera et al. reported that the monoclinic perovskite $La_2Ti_2O_7$ transforms its structure into one with the orthorhombic structure at approximately 800°C [1]. Joseph et al. predicted the nano-sized $La_2Ti_2O_7$ as a promising red phosphor for display applications due to the luminescence shown near 610 nm [2]. Ishizawa et al. reported that $La_2Ti_2O_7$ with the monoclinic structure transforms into orthorhombic structure at approximately 780°C [3]. Meng et al. have observed a narrowing of the band gap of the $La_2Ti_2O_7$ nanosheets by 0.77 eV using nitrogen doping, without any change in the particle shape, dimensions, or crystal phase [4]. Onozuka et al. have successfully prepared perovskite type $La_2Ti_2O_7$ photocatalyst with crystallinity and high surface area by sol–gel method [5]. Gao et al. have obtained the activation energy of $La_2Ti_2O_7$ as 1.34 eV and predicted that the activation energy for $La_2Ti_2O_7$ is about half of the reported band gap (2.8–3.2 eV), and thus, the DC conductivity is dominated by intrinsic charge carriers [6]. Shao et al. have proved that $La_2Ti_2O_7$ can be used as a suitable host, where the La^{3+} ions can be substituted by other trivalent lanthanide ions that induce luminescence properties in the visible region and lead to phosphor materials [7].

Most of the works in the layered perovskite $La_2Ti_2O_7$ is about their photocatalytic activity; however, the dielectric and electrical properties of $La_2Ti_2O_7$ were not performed in details. Here, we report the synthesis of $La_2Ti_2O_7$ by autoignited combustion. The structure, vibrational and optical spectroscopic studies, and the dielectric and electrical properties of the sample are also reported.

3.2 Experimental Section

Nanocrystalline $La_2Ti_2O_7$ (abbreviated as LTO) was synthesized using a modified combustion method. In this method, stoichiometric amounts of high-purity lanthanum oxide (La_2O_3) and titanium isopropoxide ($C_{12}H_{28}O_4Ti$) were dissolved in concentrated HNO_3 and methanol, respectively. To obtain the precursor complex, this stoichiometric solution was mixed with citric acid solution so that the cation ratio stays unity. The oxidant/fuel ratio of the system was adjusted using concentrated HNO_3 and liquor ammonia. The solution containing the precursor mixture at pH nearly equal to 7.0 was heated on a hot plate at around 250°C under a ventilated fume hood. The solution was boiled and undergoes dehydration, followed by decomposition leading to smooth deflation, producing a foam. On further heating, autoignition of the foam occurs due to self-propagating combustion to obtain a voluminous fluffy powder of LTO. The powder was then annealed in oxygen atmosphere at 900°C to eliminate trace organic impurity that may have remained in the sample.

Structural characterization of the as-prepared powder was performed by the powder X-ray diffraction technique using a Bruker D8 X-ray diffracto-meter with nickel-filtered Cu K_α radiation, for a step size of 0.02. The FT-IR spectrum of the sample was recorded in the range 350 to 1000 cm^{-1} with a spectrum 2, Perkin–Elmer, by the KBr pellet method. The UV-Vis spec-trum of the sample was recorded using Perkin–Elmer, Lambda 35, UV-Vis spectrometer in the range 200 to 700 nm. Photo luminescence spectrum was recorded using the JASCO, FP-8500, spectrofluorometer with SCE-846 accessory equipped with xenon lamp.

The powder was then pressed into a circular pellet of thickness 2 mm and diameter 12 mm using a hydraulic press with an average pressure of 190 MPa. The pellet was sintered at 1430°C for 2 h. The pellet was made into the form of a disc capacitor electrode by applying silver paste on both sides and preheated at 900°C for half an hour for the dielectric and impedance measurements using a Hioki 353250 LCR meter.

3.3 Results and Discussion

Figure 3.1 shows the XRD pattern of the as-prepared LTO nanopowder. All the peaks are indexed using ICDD file No. 70-1690, and the compound is found to be orthorhombic structure with space group $Pna2_1$ and Z = 8 which is in good agreement with the previous report [8]. The lattice parameters of the sample, calculated using the least square method, are a = 25.52956 Å, b = 7.89296 Å, and c = 5.57629 Å. The average crystalline size is calculated using the Debye–Scherrer formula and is found to be 31 nm.

The FTIR spectrum of LTO over the range 350–1000 cm^{-1} is shown in Figure 3.2. The band at 806.38 cm^{-1} corresponds to the characteristic vibration peak of TiO_4 tetrahedron. The absorption at 583.6 cm^{-1} is due to La–O stretching vibration, and the bands observed in the 400–500 cm^{-1} correspond to Ti–O vibrations or may be due to complex motions involving the contribution of both La and Ti cations [9, 10]. Hence, it is evident that the phase formed is of $La_2Ti_2O_7$.

The UV-Vis absorption spectra are recorded to characterize the optical absorbance of the sample. The absorption spectra of LTO are shown in

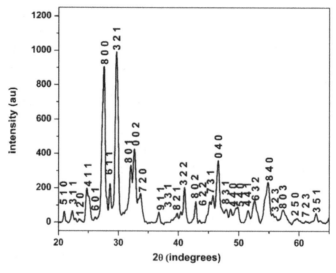

Figure 3.1   XRD pattern of LTO.

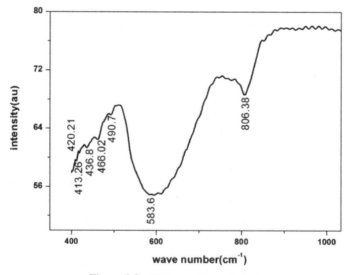

Figure 3.2   FTIR spectrum of LTO.

Figure 3.3. The optical band gap energy is determined by extrapolating the steep portion of the graph to the zero value of absorbance and solving using the relation $E = hc/\lambda$. The corresponding band gap value is 3.5 eV. The absorption and optical band gap dependence on the photon energy can be

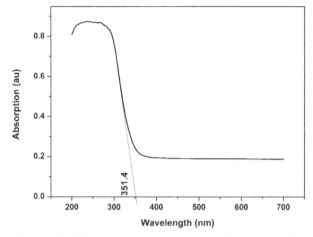

Figure 3.3   UV-Vis absorption spectrum of LTO nanopowder.

obtained by Tauc's equation:

$$\alpha h\nu = B(h\nu - E_g)^m \tag{3.1}$$

$$\alpha \text{ proportional to } F(R) = \frac{(1-R)^2}{2R} \tag{3.2}$$

where α is the absorption coefficient, B is an energy independent constant, ν is the frequency of the incident photon, F(R) is the Kubelka–Munk function, R is the reflectance (%), h is the Planck's constant, E_g is the optical band gap energy, and m is the index which depends on the nature of electronic transition responsible for the optical absorption. The value of m is 1/2 and 2 for allowed direct and indirect transitions, respectively. Tauc's plot can be extrapolated to $(\alpha h\nu)^{1/2} = 0$ to obtain the band gap [11]. Figure 3.4 shows Tauc's plots obtained from the optical absorption spectra measured at room temperature. The corresponding band gap value is 3.5 eV, which is equal to the value estimated from the absorption spectra.

The photoluminescence (PL) emission spectra of LTO under excitation at 357 nm are shown in Figure 3.5. The emission spectrum showed intense peaks in the violet and red regions. The transitions of the fundamental elements of the compounds producing these emissions are identified based on the data from Payling and Larkins [12]. The emission line observed at 409 nm may be due to the $^2F_{3.5} - {}^0\mathring{A}_{3.5}$ transition of La atom. The strong emission line observed at 433 nm may be due to the $^5F_1 - {}^3F_2^o$ transition of Ti atom. The emission at 610 nm is due to the $^2F_{3.5}^o - {}^0\mathring{A}_{2.5}$ transition of La atom.

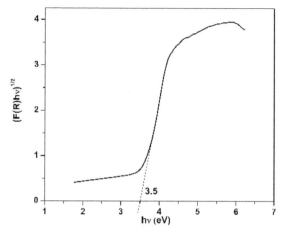

Figure 3.4   Tauc's plot for LTO nanopowder.

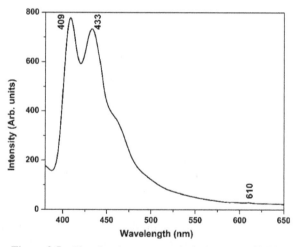

Figure 3.5   Photoluminescence emission spectra of LTO.

The strong emission intensity in the visible region shows that this compound could be beneficial for numerous optoelectronic applications such as optical amplifiers and lasers.

Dielectric constant (ε') and conductivity (σ) of LTO are studied in the frequency range 100 Hz to 5 MHz with varied temperatures ranging from 500 to 800°C. Figure 3.6 shows the variations of ε' with temperature for LTO. The value of ε' increases with temperature due to the dipolar ordering in the system under ambient temperature [13]. The high value of ε' at high

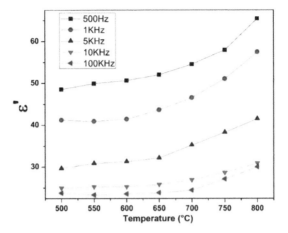

Figure 3.6 Variation of ε' with temperature.

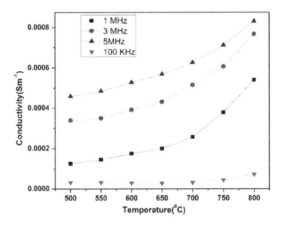

Figure 3.7 Change in conductivity with temperature.

temperature and low frequency shows that there is a free ion motion within the sample and an increase in the mobility of ions [14]. The stable value of ε' at 500°C and 1 MHz is 36.

Conductivity of the sample increases with temperature and reaches the maximum at 800°C. Figure 3.7 shows the variation of σ with temperature for the sample. Figure 3.8 shows the Arrhenius plot of the sample at 3 MHz. Activation energy calculated from the AC conductivity is 0.047 eV. The activation energy obtained for the sample at 3 MHz is below 0.8 eV, indicating the migration of oxide ions through the sample [15].

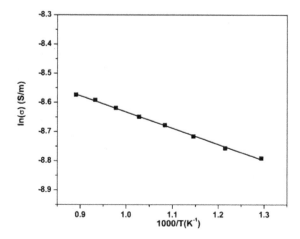

Figure 3.8   Arrhenius plot of conductivity at 3 MHz of LTO.

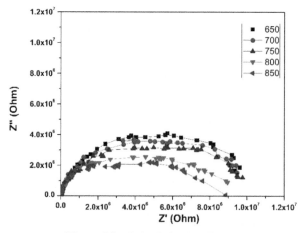

Figure 3.9   Cole–Cole plot of LTO.

Ionic conductivities of the sample is measured using impedance spectroscopy, since it can resolve the contribution of various processes such as electrode effects, bulk and interfaces, and grain boundaries in the frequency domain. Figure 3.9 shows the Cole–Cole plot of LTO at different temperatures. The curve at low frequency region bends to the real axis and becomes almost semicircular, showing increase in the conductivity of the sample. The semicircular arc in the spectrum is the combined effect of grain and grain boundary, and a small amount of electrode effect in the low frequency region

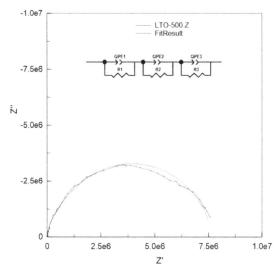

Figure 3.10   Fit result of LTO with equivalent circuit.

Table 3.1   Impedance spectra results of LTO.

T (°C)	R_g (Ω)	R_{gb} (Ω)	n_g	n_{gb}	C_g (F)	C_{gb} (F)	τ_g (s)	τ_{gb} (s)
500	922460	799490	0.825	0.938	1.11×10^{-10}	2.68×10^{-10}	1.02×10^{-04}	2.14×10^{-04}
550	912170	787310	0.865	0.974	5.82×10^{-11}	1.52×10^{-10}	5.31×10^{-05}	1.19×10^{-04}
600	900250	636470	0.925	0.989	2.88×10^{-11}	1.20×10^{-10}	2.59×10^{-05}	7.65×10^{-05}
650	702460	429490	0.932	0.999	2.53×10^{-11}	1.03×10^{-10}	1.78×10^{-05}	4.42×10^{-05}

is due to the space charge polarization. A depressed semicircle with center that lies below the positive real axis suggests deviation from ideal Debye behavior [16]. This non-Debye-type behavior of the material is due to the ion conduction among the random free energy barriers of localized ions. At high temperatures, the impedance decreases due to the relaxation time of thermally activated ions and hence the ionic conduction increases.

The grain and grain boundary microstructure can be obtained by fitting the impedance spectrum with a suitable series combination of parallel Q–R elements. Figure 3.10 shows the fitted impedance spectra for LTO at 500°C and the corresponding equivalent circuit obtained using the Z-view software within 2–3% fitting error [17]. Table 3.1 shows the values of fitted parameters. Here, R_g, and R_{gb} are the resistance of grain and grain boundary, and Q_g and Q_{gb} represent the constant phase elements of grain and grain boundary, respectively. The values of grain capacitance (C_g) and grain boundary capacitance (C_{gb}), grain relaxation time (τ_g), and grain boundary relaxation time

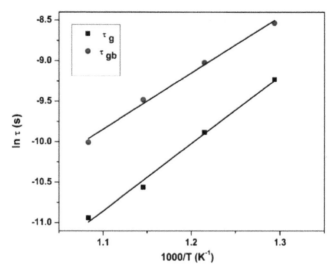

Figure 3.11   Relaxation Arrhenius plot of LTO.

(τ_{gb}) for different temperatures are also shown in Table 3.1. Constant phase element (CPE) measures the deviation of capacitance from ideal behavior, and the deviation from ideal behavior is calculated using a parameter 'n' such that:

$$C = R^{(1-n)/n}Q^{1/n} \tag{3.3}$$

The value of 'n' is unity for a pure capacitor and is zero for a pure resistor [18]. The grain and grain boundary resistance decrease with increase of temperature and the low value of R_{gb} over R_g represents the dominant behavior of grain boundary conduction over bulk conduction. The value of capacitance calculated from the fitted data shows that grain capacitance is less than the grain boundary capacitance. The relaxation time is calculated from the relation $\tau = RC$ for grain and grain boundary, which shows that the grain boundary relaxation is higher than the grain relaxation.

The activation energy is the free energy barrier that an ion has to pass through for an effective jump between the sites. The activation energies for grains and grain boundaries are calculated from the slopes of ln τ_g and ln τ_{gb} versus 1000/T plots, as shown in Figure 3.11. The obtained values of E_{ag} and E_{agb} are 0.72 eV and 0.60 eV, respectively.

3.4 Conclusions

Nanocrystalline $La_2Ti_2O_7$ ceramic was synthesized through a modified combustion technique. XRD results showed that the compound crystallized in an orthorhombic structure and had a particle size of 31 nm. Vibrational spectroscopic study supports the XRD results of the sample. The band gap energy calculated directly from the absorption spectra and from the Tauc's plot was identical. The sample showed intense emission peaks in the violet and red regions. Dielectric and impedance spectroscopic analysis of the sample showed that ions are the main source of conduction. This functional ceramic is found to be a suitable candidate for electrolyte in SOFCs, laser host material, and optoelectronic applications.

Acknowledgements

The authors acknowledge the Kerala State Council for Science, Technology, and Environment, Government of Kerala, for financial assistance.

References

[1] G. Herrera, J. Jimenez-Mier and E. Chavira, "Layered-structural monoclinic-orthorhombic perovskite $La_2Ti_2O_7$ to orthorhombic $LaTiO_3$ phase transition and their microstructure characterization," *Mater. Charact.*, vol. 89, pp. 13–22, 2014.

[2] L. K. Joseph, K. R. Dayas, S. Damodar, B. Krishnan, K. Krishnankutty, V. P. N. Nampoori and P. Radhakrishnan, "Photoluminescence studies on rare earth titanates prepared by self-propagating high temperature synthesis method," *Spectrochim. Acta Part A: Mol. Biomol. Spectrosc.*, vol. 71, no. 4, pp. 1281–1285, 2008.

[3] N. Ishizawa, F. Marumo, S. Iwai, M. Kimura and T. Kawamura, "Compounds with perovskite-type slabs. V. A high-temperature modification of $La_2Ti_2O_7$," *Acta Crystallogr. Sect. B: Struct. Crystallogr. Cryst. Chem.*, vol. 38, no. 2, pp. 368–372, 1982.

[4] F. Meng, Z. Hong, J. Arndt, M. Li, M. Zhi, F. Yang and N. Wu, "Visible light photocatalytic activity of nitrogen-doped $La_2Ti_2O_7$ nanosheets originating from band gap narrowing," *Nano Res.*, vol. 5, no. 3, pp. 213–221, 2012.

[5] K. Onozuka, Y. Kawakami, H. Imai, T. Yokoi, T. Tatsumi and J. N. Kondo, "Perovskite-type $La_2Ti_2O_7$ mesoporous photocatalyst," *J. Solid State Chem.*, vol. 192, pp. 87–92, 2012.

[6] Z. P. Gao, H. X. Yan, H. P. Ning, R. Wilson, X. Y. Wei, B. Shi, H. Ye and M. J. Reece, "Piezoelectric and dielectric properties of Ce substituted $La_2Ti_2O_7$ ceramics," *J. Eur. Ceram. Soc.*, vol. 33, no. 5, pp. 1001–1008, 2013.

[7] Z. Shao, S. Saitzek, J. F. Blach, A. Sayede, P. Roussel and R. Desfeux, "Structural characterization and photoluminescent properties of $(La_{1-x}Sm_x)_2Ti_2O_7$ solid solutions synthesized by a sol-gel route," *Eur. J. Inorg. Chem.*, no. 24, pp. 3569–3576, 2011.

[8] K. Scheunemann and H. K. Mullerbuschbaum, "Crystal-Structure of $La_2Ti_2O_7$," *J. Inorg. Nucl. Chem.*, vol. 37, no. 9, pp. 1879–1881, 1975.

[9] J. Lian, J. He, X. Zhang and F. Liu, "Fractional precipitation synthesis and photoluminescence of $La_2Ti_2O_7$:xEu^{3b} phosphors," *Solid State Sci.*, vol. 61, pp. 9–15, 2016.

[10] U. Balachandran and N. G. Eror, "X-ray diffraction and vibrational-spectroscopy study of the structure of $La_2Ti_2O_7$," vol. 4, no. 6, pp. 1525–1528, 1989.

[11] B. Vijaya Kumar, R. Velchuri, V. Rama Devi, B. Sreedhar, G. Prasad, D. Jaya Prakash, M. Kanagaraj, S. Arumugam and M. Vithal, "Preparation, characterization, magnetic susceptibility (Eu, Gd and Sm) and XPS studies of Ln_2ZrTiO_7 (Ln=La, Eu, Dy and Gd)," *J. Solid State Chem.*, vol. 184, no. 2, pp. 264–272, 2011.

[12] Richard Payling and Peter Larkins, "Optical Emission Lines of the Elements," Wiley, New York, 2000.

[13] D. G. Barton, M. Shtein, R. D. Wilson, S. L. Soled and E. Iglesia, "Structure and Electronic Properties of Solid Acids Based on Tungsten Oxide Nanostructures," *J. Phys. Chem. B.*, vol. 103, no. 510, pp. 630–640, 1999.

[14] S. G. Rathod, R. F. Bhajantri, V. Ravindrachary, P. K. Pujari and T. Sheela, "Ionic conductivity and dielectric studies of $LiClO_4$ doped poly (vinylalcohol)(PVA)/chitosan (CS) composites," *J. Adv. Dielectr.*, vol. 4, no. 4, p. 33, 2014.

[15] A. George, J. K. Thomas, A. John and S. Solomon, "Synthesis and characterization of nanocrystalline $A_6Sb_4ZrO_{18}$ (A=Ca, Sr and Ba) functional ceramics," *Solid State Ionics*, vol. 278, pp. 245–253, 2015.

[16] D. B. Dhwajam, M. B. Suresh, U. S. Hareesh, J. K. Thomas, S. Solomon and A. John, "Impedance and modulus spectroscopic studies

on $40PrTiTaO_6 + 60YTiNbO_6$ ceramic composite," *J. Mater. Sci. Mater. Electron.*, vol. 23, no. 3, pp. 653–658, 2012.

[17] M. Idrees, M. Nadeem and M. M. Hassan, "Investigation of conduction and relaxation phenomena in $LaFe_{0.9}Ni_{0.1}O_3$ by impedance spectroscopy," *J. Phys. D: Appl. Phys.*, vol. 43, no. 15, p. 155401, 2010.

[18] J. Ross Macdonald, "Note on the parameterization of the constant-phase admittance element," *Solid State Ionics*, vol. 13, no. 2, pp. 147–149, 1984.

4

Structure and Optical Properties of Y_6WO_{12}:0.005 Sm^{3+}

N. L. Jayalekshmy, Jijimon K. Thomas and Sam Solomon[*]

Department of Physics, Mar Ivanios College, Thiruvananthapuram 695015, Kerala, India
E-mail: samdmrl@yahoo.com
*Corresponding Author

Cubic Y_6WO_{12} nanoparticles are synthesized by a combustion method. The prepared materials are characterised by X-ray diffraction (XRD), Fourier Transform Infrared Spectroscopy (FT-IR), UV-visible Spectroscopy and Photoluminescence Spectroscopy (PL). The X-ray diffraction studies show that Y_6WO_{12} crystallize in cubic structure. The average crystalline size calculated by Scherrer's method is about 23 nm. The lattice parameters obtained from the XRD analysis are in good agreement with the reference data. The IR spectrum shows the band corresponding to Y-O, W-O metal oxygen bond vibrations. The Tauc's plot shows that it is an indirect bandgap material with a band gap 3.36 eV. The UV visible spectra of Y_6WO_{12}:0.005 Sm^{3+} shows absorption peaks of Sm^{3+}. A blue emission of Sm^{3+} nearly 458 nm is observed in the PL spectrum of this phosphor.

4.1 Introduction

Nanomaterials have gained technological importance due to their physical and chemical properties arising from their morphology, size and dimension. Nanomaterials have very promising properties due to large ratio of surface area to volume [1]. Among the nanomaterials the rare earth compounds have been widely used in the field of catalysts, luminescent devices based on their electronic, optical and chemical characteristics arising from their 4f electrons.

The properties of nanomaterials can be tuned by controlling the size [2]. The structure of Cubic Y_6WO_{12} is a defect fluorite-type structure. Fluorite related oxides have properties such as catalytic activity, transformation toughening and various phase transformations [3]. The various stoichiometric compositions of rare earth tungsten double oxides with different mole ratios have been studied. Borchardt reported the stoichiometries of Y_2O_3:WO_3 = 1:3, 1:1, 15:8, 9:4 and 3:1 as stable phases and 3:2 (X phase) as a metastable phase [4]. The melting temperature of Y_6WO_{12} is very high and is about $2360 \pm 20°C$. The structure of these compound is similar to Y_6UO_{12} and contain WO_6 groups [5]. The photoluminescense property, catalytic activity and ionic conductivity of Y_6WO_{12} have investigated earlier. The WO_6 groups in the host lattice can absorb UV light and can transfer energy to activator ions [6].

A variety of synthesis methods are reported in the literature for the synthesis of Y_6WO_{12} [4–6]. In the present work a single step combustion method is used for the preparation of Y_6WO_{12} nanomaterial. Combustion technique is an important method used for the preparation of nanomaterials which involves the exothermic reaction of an oxidizer and a fuel.

4.2 Experimental

Y_6WO_{12} nanoparticles are prepared through combustion technique using the corresponding metal nitrate (oxidizing agent) and suitable fuel (reducing agent) [8]. The starting materials are yttrium oxide (Y_2O_3, 99.9%), and Ammonium tungsten oxide hydrate$(NH_4)_6W_{12}O_{39}·xH_2O$. The rare earth oxides is dissolved in nitric acid and ammonium tungstate in distilled water. Urea (NH_2CONH_2) is added as fuel reagent (reducing agent) and citric acid is used as complexing agent to form the precursor complex [9]. Stoichiometric amounts of oxidizing and reducing agents are dissolved in a minimum volume of deionized water to obtain transparent aqueous solution in a glass beaker, which is subsequently heated using a hot plate at $250°C$ in a ventilated fume hood. The solution boils on heating and undergoes dehydration accompanied by foaming. On persistent heating, the foam gets auto ignited, due to self-propagating combustion giving a voluminous fluffy powder. The powder obtained after auto-ignition is annealed at $800°C$ for one hour to obtain pure, nanocrystalline powder. The prepared sample is characterized by using X-ray diffractometer (D8 advance, Bruker, Germany) with CuK_α radiation in the range of 20–70 in steps of 0.02. The infrared spectrum of the sample is recorded in the range 400–1000 cm^{-1} on a Fourier Transform Infrared

(FTIR) Spectrometer (Spectrum2, Perkin-Elmer, Singapore) using the ATR method. The diffuse reflectance spectrum of the sample is recorded in the range 200 to 800 nm using a UV-Vis spectrometer (Lambda 35, Perkin-Elmer, Singapore) with an integrated sphere accessory (RSA-PE-20, Lab sphere, USA). The Photoluminescence spectrum of the sample is recorded using JASCO, FP-Spectro fluorometer with SCE-846 accessory.

4.3 Results and Discussions

4.3.1 XRD

XRD is an analytical technique used to investigate the phase, structure and average crystallite size. The XRD of as prepared Y_6WO_{12} nanoparticles is shown in the Figure 4.1. It confirms the formation of single phase cubic Y_6WO_{12} nanoparticles without any impurity peaks [10]. All the peaks indexed matches well with JCPDS No. 35-0174. The broadening of the peaks in the XRD pattern is due to nanoscale of materials. The particle size of prepared nanoparticles is calculated from Full Width Half Maximum (FWHM) of the diffraction peaks using Scherrer Equation

$$D = K\lambda/\beta \, \mathrm{Cos} \, \theta$$

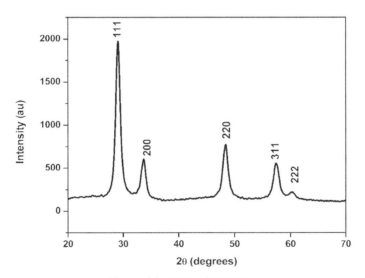

Figure 4.1 XRD of Y_6WO_{12}.

where D is the crystal size, λ is the $CuK_{\alpha 1}$ wavelength (1.5406 Å), β is the full width half maximum of the peak in radian and θ is the corresponding diffraction angle. The average particle size calculated by this method is about 23.08 nm.

The lattice constants for the cubic structure are calculated using the equation

$$1/d^2 = 1/a^2(h^2 + k^2 + l^2)$$

The value of the calculated lattice constant is a = 5.321 Å and it agrees well with the reference data.

4.3.2 FTIR

The vibrational structure of sample is analyzed by Fourier Transform Infrared (FTIR) spectrometer in the range 400–1000 cm^{-1} with a resolution of 4 cm^{-1} is shown in the Figure 4.2. FTIR spectroscopy is widely used to identify the functional groups and vibrational structure of atoms and molecules. The spectrum shows the metal oxygen vibrations of Y-O and W-O. The bands below 650 cm^{-1} are assigned to Y-O vibrational modes and 400 cm^{-1} to 1000 cm^{-1} are also due to W-O stretching and W-O-W bridging stretching modes. The FTIR shows all vibrational modes which are in agreement with the literature [11–13].

4.3.3 Optical Properties

4.3.3.1 UV-visible spectroscopy

UV-visible spectroscopy is a very powerful tool for characterizing optical and electronic properties of as prepared nanoparticles. It measures the percentage of radiation in the ultra violet (200–400 nm) and visible (400–800 nm) regions that is absorbed at each wavelength. The absorption spectrum of the sample measured in the range 200 to 800 nm is shown in the Figure 4.3. From the figure it is clear that sample absorbs heavily in the UV region. The absorption band from 200–350 nm is due to charge transfer bands (CTB), from oxygen to tungsten (O^{2-} to W^{6+}) [14]. The optical absorption band edge value can be determined by extrapolating the absorption onset in the UV region to X-axis. The intersecting wavelength is then converted to corresponding energy using $E = hc/\lambda$.

The absorption band edge wavelength is 365 nm and corresponding energy is 3.4 eV.

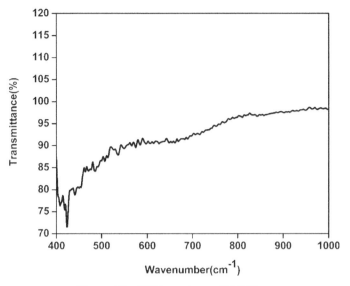

Figure 4.2 FTIR spectrum of Y_6WO_{12}.

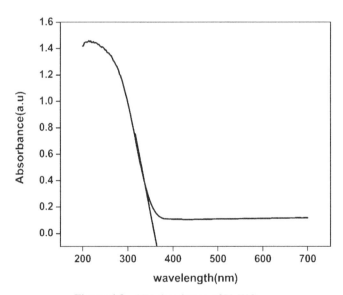

Figure 4.3 UV absorbance of Y_6WO_{12}.

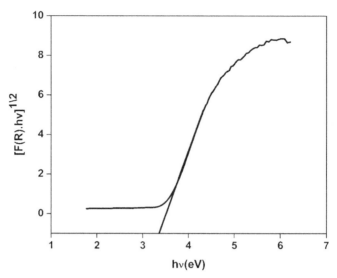

Figure 4.4   Band gap of Y_6WO_{12}.

The dependence of absorption co-efficient on photon energy is calculated using Tauc's equation

$$(\alpha h\nu) = B(h\nu - E_g)^m$$

Where α is the absorption co-efficient, h is the Planck's constant, ν is the frequency of incident radiation, B is the energy independent constant, E_g is the optical band gap energy and m is an index which depends on the nature of electronic transition responsible for the optical absorption. The value of m is 1/2 and 2 for allowed direct and indirect transitions respectively. The band gap can be obtained by extrapolating the straight-line portion of the plot to x axis (Figure 4.4). Y_6WO_{12} is identified as an indirect band gap material and its optical band gap obtained by Tauc's equation is (3.3). The optical absorption and band gap of Y_6WO_{12}:0.005 Sm^{3+} is shown in the Figures 4.5 and 4.6 respectively. A peak at 410 nm is seen in the absorption spectrum, which is due to $^6H_{5/2}$ to $^4K_{11/2}$ transition of Sm^{3+} [14]. The absorption wavelength obtained also is 366 nm and the corresponding energy is 3.38 eV. The band gap of Y_6WO_{12}:0.005 Sm^{3+} obtained from Tauc's plot is about 3.5 eV.

4.3.3.2 Photoluminescence

Rare earth doped inorganic nanomaterials show luminescence because the 4f orbital lies inside and is strongly shielded from the outermost filled $5s^2$

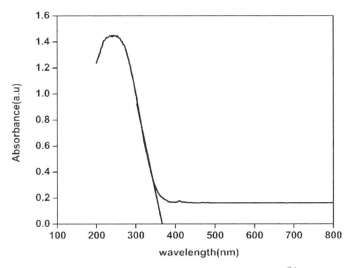

Figure 4.5 UV absorbance of Y_6WO_{12}:.005 Sm^{3+}.

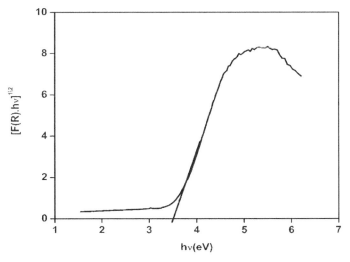

Figure 4.6 Band gap of Y_6WO_{12}:.005 Sm^{3+}.

and $5p^6$ orbitals. Sm^{3+} ion is one of the promising activator among the trivalent rare earth ions owing to the efficient luminescence in the visible region. So, it can be used in the optical devices [15]. Figures 4.7(a) and 4.7(b) deals with the photoluminescence spectra of the compound. Sm^{3+}

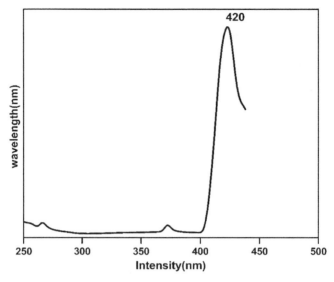

Figure 4.7(a)   Photoluminescence excitation spectra of Y_6WO_{12}:0.005 Sm^{3+}.

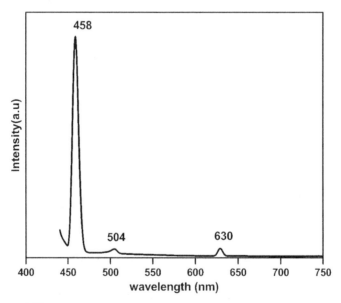

Figure 4.7(b)   Photoluminescence emission spectra of Y_6WO_{12}:0.005 Sm^{3+}.

doped materials have various application in the WLED fluorescent devices and colour displays [16, 17]. When excited at 420 nm three emissions are observed at 458, 504 and 630 nm which are assigned to $^6H_{5/2} \rightarrow {}^4I_{11/2}$ (458 nm), $^4G_{5/2} \rightarrow {}^6H_{5/2}$ (504 nm), $^4G_{5/2} \rightarrow {}^6H_{7/2}$ (630 nm) [18, 19].

4.4 Conclusions

Y_6WO_{12} nanoparticles are prepared through combustion synthesis. The XRD results show that it has a particle size of 23.08 nm and its structure is cubic defect fluorite. FTIR shows metal oxide bond vibrations of Y-O and W-O. From the UV visible studies, it is an indirect gap material with a band gap of 3.36 eV and the band gap of Y_6WO_{12}:.005 Sm^{3+} is 3.5 eV. A small shift is observed in the band gap when Y_6WO_{12} is doped with .5% of Sm^{3+}. Photoluminescence studies of Y_6WO_{12}:.005 Sm^{3+} shows absorptions peaks of Sm^{3+}.

Acknowledgements

The authors acknowledge the Kerala State Council for Science, Technology and Environment, Government of Kerala for financial assistance.

References

[1] T. Liu, Y. Zhang, H. Shao, X. Li, Synthesis and characteristics of Sm_2O_3 and Nd_2O_3 nanoparticles, Langmuir. 19 (2003) 7569–7572. doi:10.1021/la0343501.

[2] F. Lei, B. Yan, H. H. Chen, Solid-state synthesis, characterization and luminescent properties of Eu^{3+}-doped gadolinium tungstate and molybdate phosphors: Gd $(2-x)MO_6$:Eux^{3+} (M = W, Mo), J. Solid State Chem. 181 (2008) 2845–2851. doi:10.1016/j.jssc.2008.07.008.

[3] M. Yoshimura, J. Ma, M. Kakihana, Low-Temperature Synthesis of Cubic and Rhombohedral Y_6WO_{12} by a Polymerized Complex Method, J. Am. Ceram. Soc. 81 (1998) 2721–2724. doi:10.1111/j.1151-2916.1998.tb02684.x.

[4] K. Kuribayashi, M. Yoshimura, T. Ohta, T. Sata, High-Temperature Phase Relations in the System Y_2O_3-$Y_2O_3 \cdot WO_3$, J. Am. Ceram. Soc. 63 (1980) 644–647. doi:10.1111/j.1151-2916.1980.tb09853.x.

[5] T. Chien, J. Yang, C. Hwang, M. Yoshimura, Synthesis and photoluminescence properties of red-emitting Y_6WO_{12}:Eu^{3+} phosphors, J. Alloys Compd. 676 (2016) 286–291. doi:10.1016/j.jallcom.2016.03.083.

[6] R. Yu, D. S. Shin, K. Jang, Y. Guo, H. M. Noh, B. K. Moon, B. C. Choi, J. H. Jeong, S. S. Yi, Photoluminescence Properties of Novel Host-Sensitized Y_6WO_{12}:Dy^{3+} Phosphors, J. Am. Ceram. Soc. 97 (2014) 2170–2176. doi:10.1111/jace.12937.

[7] L. K. Bharat, Y. Il Jeon, J. Su, Citrate-based sol – gel synthesis and luminescent properties of Y_6WO_{12}:Eu^{3+}, Dy^{3+} phosphors for solid-state lighting applications, Ceram. Int. (2016) 1–9. doi:10.1016/j.ceramint.2015.12.093.

[8] A. K. Tyagi, Combustion Synthesis: a Soft-Chemical Route for Functional Nano-Ceramics, Chem. Res. Soc. (2007) 39–48.

[9] K. C. Patil, S. T. Aruna, T. Mimani, Combustion synthesis: an update, Curr. Opin. Solid State Mater. Sci. 6 (2002) 507–512. doi:10.1016/S1359-0286(02)00123-7.

[10] T. Ishigaki, N. Matsushita, M. Yoshimura, K. Uematsu, K. Toda, Keywords: melt synthesis, tungstate, La_2WO_6, red phosphor, Phys. Procedia. 2 (2009) 587–601. doi:10.1016/j.phpro.2009.07.045.

[11] M. R. Davolos, Ã. S. Feliciano, A. M. Pires, R. F. C. Marques, M. J. Jr, Solvothermal method to obtain europium-doped yttrium oxide, 171 (2003) 268–272. doi:10.1016/S0022-4596(02)00174-3.

[12] R. Srinivasan, R. Yogamalar, A. C. Bose, Structural and optical studies of yttrium oxide nanoparticles synthesized by co-precipitation method, Mater. Res. Bull. 45 (2010) 1165–1170. doi:10.1016/j.materresbull.2010.05.020.

[13] J. Yu, J. Xiong, B. Cheng, Y. Yu, J. Wang, Hydrothermal preparation and visible-light photocatalytic activity of Bi_2WO_6 powders, 178 (2005) 1968–1972. doi:10.1016/j.jssc.2005.04.003.

[14] G. Annadurai, S. M. M. Kennedy, V. Sivakumar, Photoluminescence properties of a novel orange-red emitting $Ba_2CaZn_2Si_6O_{17}$:Sm^{3+} phosphor, J. Rare Earths. 34 (2016) 576–582. doi:10.1016/S1002-0721(16)60064-9.

[15] F. Cheviré, F. Clabau, O. Larcher, E. Orhan, F. Tessier, R. Marchand, Tunability of the optical properties in the $Y6(W,Mo)(O,N)_{12}$ system, Solid State Sci. 11 (2009) 533–536. doi:10.1016/j.solidstatesciences.2008.06.009.

[16] Y.-H. Li, J.-F. Huang, J.-Y. Li, L.-Y. Cao, J. Lu, J.-P. Wu, A hydrothermal assisted method to prepare Samarium Tungstate sheets

at lowered reaction temperature, Mater. Lett. 135 (2014) 168–171. doi:10.1016/j.matlet.2014.07.164.

[17] F. Yang, Z. Yang, Q. Yu, Y. Liu, X. Li, F. Lu, Sm^{3+}-doped $Ba_3Bi(PO_4)_3$ orange reddish emitting phosphor, Spectrochim. Acta – Part A Mol. Biomol. Spectrosc. 105 (2013) 626–631. doi:10.1016/j.saa.2013.01.010.

[18] J. Zhang, X. Wu, J. Zhu, Q. Ren, Luminescence properties of a novel $CaLa_4Si_3O_{13}$:Sm^{3+} phosphor for white light emitting diodes, Opt. Commun. 332 (2014) 223–226. doi:10.1016/j.optcom.2014.07.013.

[19] P. Van Do, V. Phi, V. Xuan, N. Trong, V. Thi, T. Ha, N. M. Khaidukov, Y. Lee, B. T. Huy, Judd – Ofelt analysis of spectroscopic properties of Sm^{3+} ions in K_2YF_5 crystal, J. Alloys Compd. 520 (2012) 262–265. doi:10.1016/j.jallcom.2012.01.037.

MODULE 2

5

UV-Visible Spectroscopy: An Attempt in Determining Dispersion Characteristics of Carbon Black-Filled Poly (Lactic Acid) Composites

Ajay Vasudeo Rane[1,*], Neehara Alackal[2] and Krishnan Kanny[1,*]

[1]Composite Research Group, Department of Mechanical Engineering, Durban University of Technology, Durban 4000, South Africa
[2]Department of Chemistry Deva Matha College, Mahatma Gandhi University, Kottayam, Kuravilangad, Kottayam 686633, Kerala, India
E-mail: ranea061@gmail.com; kannyk@dut.ac.za
*Corresponding Authors

Spectroscopic techniques are increasingly applied in investigations on high molecular weight compounds. Absorption spectra provide a substantial amount of information on the relative arrangement of the structural elements in the system of interest and on the interactions, operating between these elements, which hold them in a definite spatial configuration. Moreover, the spectra reflect the stoichiometric relationships among individual structural elements in the molecule. In this study, we have fabricated neat poly (lactic acid) and carbon black-filled poly (lactic acid) films via dissolution–dispersion method and determined their absorption characteristics in the range 250–800 nm, as an attempt to determine dispersion characteristics and adlayers in carbon black filled poly (lactic acid) composites.

5.1 Introduction

Analytical techniques are one of the most powerful tools available for the study of atomic and molecular structure and are used in the analysis of most of the samples. Analytical technique or spectroscopy deals with the

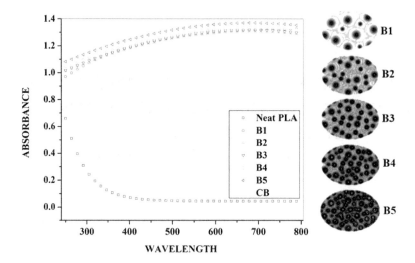

study of interaction of electromagnetic radiation with the matter. During the interaction, the energy is absorbed or emitted by the matter [1]. The measurement of this radiation frequency (absorbed or emitted) is made using spectroscopy. The study of spectroscopy can be carried out under the following headings: atomic spectroscopy and molecular spectroscopy. Atomic spectroscopy deals with the interaction of electromagnetic radiation with atoms, during which the atoms absorb radiation and get excited from ground state electronic energy level to another. Molecular spectroscopy deals with the interaction of electromagnetic radiation with the molecules, and this results in transition between rotational, vibrational and electronic energy levels [1]. The mechanism underlying absorption spectra may be visualized as follows: if electromagnetic radiation falls on an assembly of molecules, as a result of interaction of the radiation with these molecules, some of the radiation quanta are absorbed. This attenuation of the original radiation intensity is recorded as the absorption spectrum by a spectrophotometer, that is, the molecule absorbs photon of energy and undergoes a transition from the lower energy level to the higher energy level; the measurement of this decrease in the intensity of radiation is the basis of absorption spectroscopy; and the spectrum obtained is called absorption spectrum [2]. If the molecules come down from the excited state to the ground state with the emission of photons of energy, the spectrum is called emission spectrum [1]. The absorption spectra related to the changes in the rotational energy alone occur in the far infrared region >100 µm, vibration–rotation spectra fall in the classical infrared region (2–25 µm), while the changes in the electronic energy of the molecule give rise to absorption in the visible and ultraviolet parts of the electromagnetic spectrum. For investigations of polymers, the most common spectral range is the infrared, together with near infrared region. Studies in ultraviolet and visible parts of the spectrum are of less importance due to its inability in structural determination. Herein we propose UV-visible region within the range 250–800 nm to determine the dispersion of a filler within the polymeric matrix.

Fillers have played a major role in meeting these ever more demanding requirements. The introduction of plastic components in automobiles has been of great value. Early attempts to use plastics failed because they lacked strength and weather resistance. Fillers have been responsible for transforming these same plastics to strong durable automotive components. Portable computer has become the truly portable laptop of today due in large part to the lighter, strongly reinforced plastics that are now available. The cases not only look smooth and sleek, and they provide shielding from the electromagnetic radiation that is used to prevent the use of computers on aircraft in flight. Where filler used to be thought of as a means to lower cost of a plastic part, they now contribute to the unique properties that sophisticated users demand. In fact, many fillers now cost more than the polymers that they are added to [3]. But such additions make economic sense because of the value that the filler brings to the formulation. Many fillers can be used to influence chemical reactions occurring in their presence. The reaction rate can be decreased or increased. Fillers such as zinc oxide will react with UV degradation products in polyethylene to limit damage. The pot-life of curing mixtures can be increased. Cure rates can be slowed, exothermic effects can be controlled, incompatible polymers can be blended, and molecular mobility reduced [3].

The work presented through this manuscript is at its initial stages, more details are yet to be analyzed; through this manuscript, we propose and attempt a method to determine the dispersion characteristics of filler in a polymer matrix (carbon black particles in poly [lactic acid]) and to an extent the formation and strength of adlayers.

5.2 Experimental Section

5.2.1 Materials

Poly (lactic acid) (PLA)—biopolymer 3100HP, procured from Nature works USA, was used in our study. Carbon black—N220, with surface area, DBP number, and density of 100–120 m^2/g, 113 cm^3/100 g, and 1.8 g/cm^3, respectively, procured from CABOT Ltd, was used as an active filler with varied loadings, ranging from 0.5 (B1), 1.0 (B2), 1.5 (B3), 2.0 (B4), and 2.5 (B5) weight percent.

5.2.2 Processing Information for Poly (Lactic Acid) and Carbon Black-Filled Poly (Lactic Acid) Composites Via Dissolution–Dispersion Technique on a Laboratory Scale [5]

Poly (lactic acid) granules were dried in hot air oven to remove moisture (ensure constant weight to confirm complete removal of moisture). Oven-dried poly (lactic acid) was added to chloroform and kept overnight for the dissolution of poly (lactic acid) to take place effectively, ensure the reactor is well sealed to avoid evaporation of chloroform and, in addition to dissolution process, confirm maximum solvation of poly (lactic acid) chains sonication on a water bath sonicator. Carbon black dispersions were prepared using chloroform in a water bath sonicator at room temperature. Thereafter, the prepared carbon black dispersions were mixed with poly (lactic acid) dissolution and kept overnight to ensure the distribution of carbon black aggregates in poly (lactic acid) dissolution. In order to ensure maximum interaction between poly (lactic acid) chains and carbon black aggregates, the poly (lactic acid)/carbon black solution was sonicated on a water bath sonicator. Further, carbon black/poly (lactic acid) chloroform solution was casted on a Teflon plate and dried in vacuum oven for removal of chloroform, in order to fabricate a sheet of poly (lactic acid) and poly (lactic acid)/carbon black composites.

5.2.3 Characterization: UV-Visible Absorbance Spectra

All organic compounds absorb ultraviolet light, although in some instances of very short wavelength. For practical reasons, we shall be concerned with absorbance above 200 nm; from the available literatures, an absorbance peak at 230 nm is observed for PLA [4], and hence in our study, we have performed UV-visible analysis in the range 250–800 nm, thereby neglecting the transition region and considering the wavelength thereafter shall be helpful in determining the dispersion characteristics of carbon black into poly (lactic acid) matrix.

5.3 Results and Discussion

A combined graph of absorbance values from 250 to 800 nm for neat PLA, carbon black and carbon black filled poly (lactic acid) specimens are presented in beginning of the chapter. To have a better understanding of absorption values at different wavelengths of UV-Vis spectra, graphs are being plotted from 250 to 400 nm for UV spectra, to study absorbance with increasing loading of carbon black and also to study the absorbance at different wavelengths for a particular loading of carbon black. Similar plots are made for visible region from 400 to 800 nm.

5.3.1 Absorbance Spectra for Ultraviolet Range from 250 to 400 nm

A difference in the absorbance values for PLA and carbon black in the ultraviolet region can be observed in Figure 5.1. Addition of carbon black to poly (lactic acid) increases the absorption characteristics of carbon black and imparts absorption property (absorption of UV light) to poly (lactic acid), thereby confirming the synergism between carbon black and poly (lactic acid) and can be observed in Figures 5.2–5.9. The absorbance values increase as

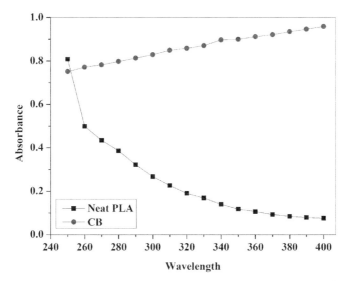

Figure 5.1 Absorbance value in UV region for neat PLA and carbon black (CB).

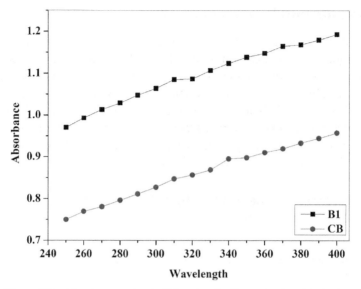

Figure 5.2   Absorbance value in UV region for B1 and carbon black (CB).

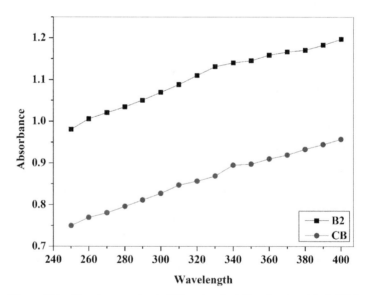

Figure 5.3   Absorbance value in UV region for B2 and carbon black (CB).

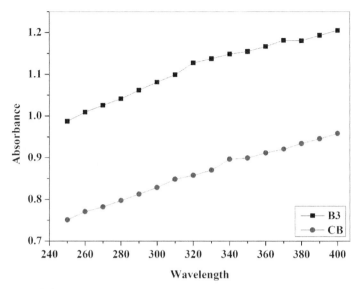

Figure 5.4   Absorbance value in UV region for B3 and carbon black (CB).

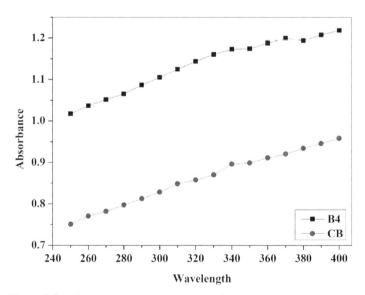

Figure 5.5   Absorbance value in UV region for B4 and carbon black (CB).

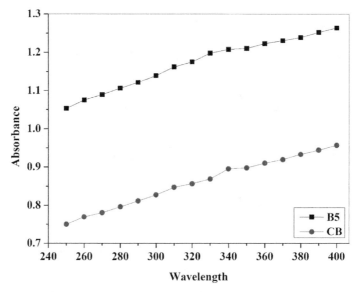

Figure 5.6   Absorbance value in UV region for B5 and carbon black (CB).

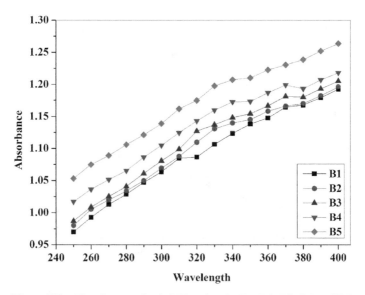

Figure 5.7   Absorbance value in UV region for B1, B2, B3, B4, and B5.

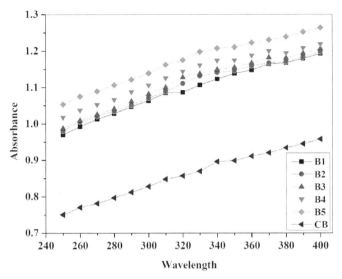

Figure 5.8 Absorbance value in UV region for B1, B2, B3, B4, B5, and carbon black (CB).

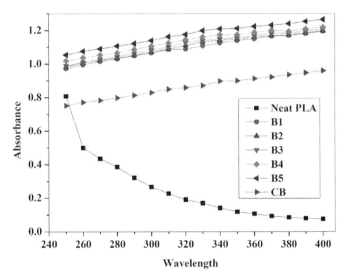

Figure 5.9 Absorbance value in UV region for neat PLA, B1, B2, B3, B4, B5, and carbon black (CB).

the loading of carbon black in PLA increases, which can be confirmed by the graph in Figure 5.8. Absorbance values for carbon black-filled PLA composites increased with the increase in concentration of carbon black, which can be attributed to existence of sp^2 carbons in carbon black [5]. Considering the internal structure of carbon black filled poly (lactic acid) composites, increase in number of three-dimensional networks formed by carbon black within poly (lactic acid) matrix may contribute to the absorbance, owing to increase in number of adsorption sites of poly (lactic acid) chains on surface of carbon black. Absorbance values for each concentration of carbon black increased with increase in wavelength in the ultraviolet region, which can be attributed to "near to homogenous dispersion" [6] of carbon black and formation of strong adlayers within poly (lactic acid) matrix.

5.3.2 Absorbance Spectra for Visible Range from 400 to 800 nm

Similar trends in absorbance is observed for carbon black filled poly (lactic acid) composites in visible region of the spectrum. A considerable difference in absorbance for PLA and carbon black in visible region is observed as shown in Figure 5.10. Addition of carbon black within poly (lactic acid) matrix increases the absorption limit of carbon black, also imparts absorption property (absorption of visible light) to poly (lactic acid), thereby confirms the existence of synergistic effect between poly (lactic acid) and carbon black which is clearly observed in Figures 5.11–5.18.

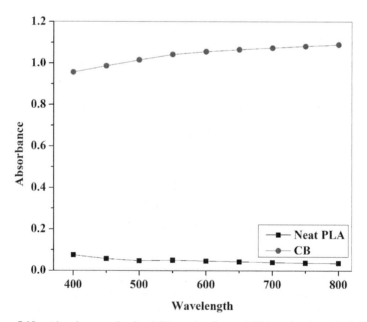

Figure 5.10   Absorbance value in visible region for neat PLA and carbon black (CB).

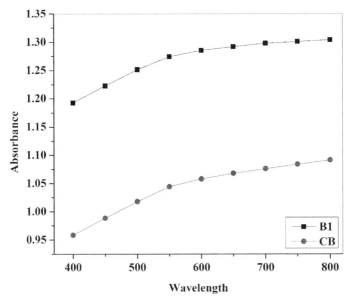

Figure 5.11 Absorbance value in visible region for B1 and carbon black (CB).

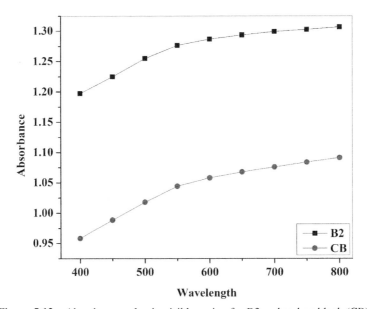

Figure 5.12 Absorbance value in visible region for B2 and carbon black (CB).

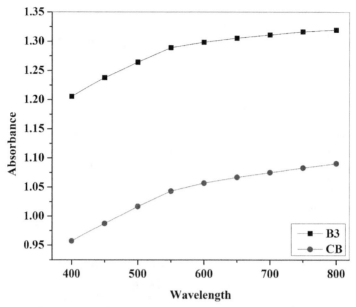

Figure 5.13   Absorbance value in visible region for B3 and carbon black (CB).

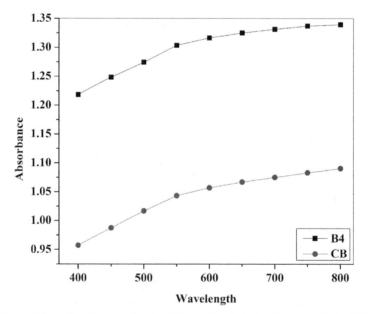

Figure 5.14   Absorbance value in visible region for B4 and carbon black (CB).

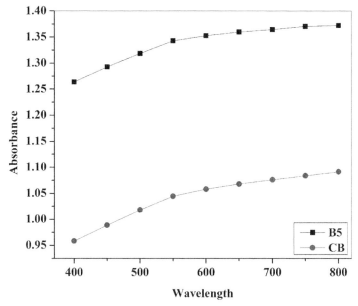

Figure 5.15 Absorbance value in visible region for B5 and carbon black (CB).

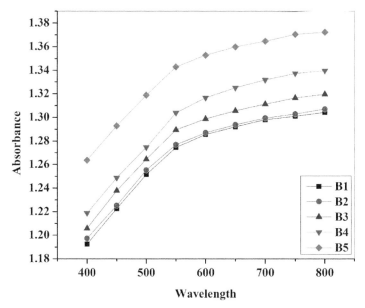

Figure 5.16 Absorbance value in visible region for B1, B2, B3, B4, and B5.

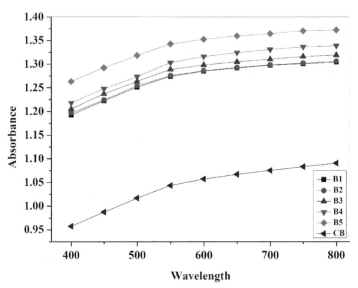

Figure 5.17 Absorbance value in visible region for B1, B2, B3, B4, B5, and carbon black (CB).

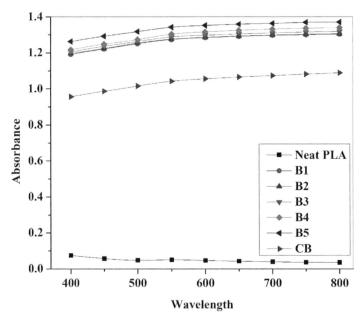

Figure 5.18 Absorbance value in visible region for neat PLA, B1, B2, B3, B4, B5, and carbon black (CB).

It is observed that, absorbance increase as the loading of carbon black in PLA increases, which can be confirmed by plot in Figure 5.18. Absorbance values for carbon black-filled PLA composites increased with the increase in concentration of carbon black, which can be ascribed to association of sp^2 carbons with carbon black [5]. On keen observation, taking in to consideration the structural features within poly (lactic acid) composites, three-dimensional networks of carbon black within poly (lactic acid) matrix shall contribute to increase in absorption limit. Adsorption of poly (lactic acid) chains on surface of carbon black enhances the absorbance limit. Increase in carbon black, increases the probability of adsorption sites formation and enhances the absorbance in visible region. Further absorbance for each concentration of carbon black increased with increase in wavelength in the visible region, which can be attributed to formation of a "near to homogenous dispersion" [6] of carbon black and formation of strong adlayers within poly (lactic acid) matrix.

5.4 Conclusion

The increase in absorbance with the increase in wavelength confirms near to homogenous dispersion of carbon black and formation of strengthened adlayers within the carbon black filled poly (lactic acid) composites. The increase in absorbance limit with increasing carbon black content in poly (lactic acid) composites can be attributed to presence of sp^2 carbons and number of adsorption sites formed within carbon black filled poly (lactic acid) composites. Hence through this study, an attempt to determine the level of dispersion of carbon black particles and to an extent determine the formation and strength of adlayers within poly (lactic acid) matrix via UV-visible spectroscopy based on wavelength vs absorbance plots was analyzed. Finally through observations made by our experiments, UV-visible spectroscopy can be used as a qualitative tool to determine the structural features within polymer composites using absorbance values from UV–Visible spectra.

Acknowledgement

The author (Ajay Vasudeo Rane) is thankful to CSIR, South Africa, for their CSIR-DST Inter Programme Bursary Award (2016, 2017 and 2018) in the area of composites. The authors also extend their thanks to IUIC-Kottayam,

Mahatma Gandhi University, India, and Durban University of Technology, South Africa, for their facilities to carry out this work.

References

[1] A. Ravikrishnan (2008). Engineering chemistry-I, Shri Krishna Publications, India.

[2] J. Urbanski, W. Czerwinski, K. Janicka, F. Majewska and H. Zowall (1977). Handbook of analysis of synthetic polymers and plastics. Ellis Horwood Limited Publishers, Chichester.

[3] G. Wypych (2000). Handbook of fillers, 2nd Edition, ChemTec Publishing, Toronto, NY.

[4] G. Costanzo, L. Ribba, S. Goyanes and S. Ledesma (2014). Enhancement of the optical response in a biodegradable polymer/azo-dye film by the addition of carbon nanotubes. Journal of Physics D Applied Physics. 47. doi:10.1088/0022-3727/47/13/135103

[5] B.P. Grady (2011). Carbon Nanotube-Polymer Composites Manufacture, Properties, and Applications, Hoboken, New Jersey.

[6] G. Wypych (2012). Atlas of Material Damage, 1st Edition, ChemTech Publishing, Toronto, NY.

6

Eco-Friendly Synthesis of Silver Nanoparticles by Using *Moringa oleifera* Leaf Extract as Reducing Agent and Their Catalytic Activity with MB Dye

Syeda Anjum Mobeen, Maram Vidya Vani and Khateef Riazunnisa[*]

Department of Biotechnology and Bioinformatics, Yogi Vemana University, Kadapa, Andhra Pradesh, India
E-mail: khateefriaz@gmail.com; krbtbi@yogivemanauniversity.ac.in
[*]Corresponding Author

The development of biologically inspired experimental processes for the synthesis of nanoparticles is evolving into an important branch of nanotechnology. This study reports the synthesis of silver nanoparticles (AgNPs) using *Moringa oleifera* leaves of aqueous solution as a potential reducing as well as stabilizing agent with the help of 8 mM silver nitrate solution at 2.5% leaf concentration. It also examines phytochemical screening to excavate the most important constituents like phenols, carbohydrates, proteins, tannins, alkaloids, flavonoids, steroids, glycosides, and diterpenes, degradation of methylene blue (MB) dye using biologically synthesized AgNPs. The aqueous leaves extract was added to silver nitrate solution; the color of the reaction medium was changed from pale yellow to brown that indicates reduction of silver ions to AgNPs. Thus, synthesized AgNPs were characterized by UV-Vis spectrophotometer that showed an absorption peak at 460 nm. The photocatalytic activity of the synthesized AgNPs was observed to have potential efficacy to degrade methylene blue (MB) dye at room conditions. This was confirmed by the decrease in maximum absorbance of MB dye with respect to time using UV-Vis spectrophotometer.

The bio-synthesized AgNPs effectively degraded nearly 73.11% MB with *M. oleifera* AgNPs at 4 hours of exposure time.

6.1 Introduction

Nanotechnology deals with the synthesis of nanoparticles with controlled size, shape, and dispersity of materials at the nanometer scale length [1] and their potential use for human well-being. Among all nanoparticles, noble metal nanoparticles have enormous applications in diverse areas such as bioimaging, sensor, diagnosis, and novel therapeutic in biomedical field [2]. Metallic silver and silver nanoparticles (AgNPs) were recently applied as antimicrobial agents in various products such as cosmetics [3], animal feed [4], coating of catheters [5], wound dressing [6], and water purification [7] with a minimal risk of toxicity in humans. Silver (Ag), a noble metal, has potential applications in medicine due to its unique properties such as good conductivity, chemically stable catalytic activity, surface-enhanced raman scattering, and antimicrobial activity. It increases the oral bio-availability and overcomes the poorly water soluble herbal medicines [8–10]. Green synthesis of nanoparticles is an emerging branch of nanotechnology. Biosynthesis of nanoparticles using plant extracts is the favorite method of green, exploited to a vast extent because the plants are widely distributed, easily available, advancement over physical and chemical methods, safe to handle and with a range of metabolites and compatibility for pharmaceutical and biomedical applications as they do not use toxic chemicals in the synthesis protocols [11–14]. In recent years, there has been an increasing demand for AgNPs due to their unique physicochemical properties such as catalysis, magnetic and optical polarizability, electrical conductivity, and surface-enhanced raman scattering. These properties have been widely utilized in photography, catalysis, biological labeling, photonics, optoelectronics, and information storage.

Moringa oleifera Lam. belongs to the family Moringaceae. It is a fast growing deciduous shrub or small tree up to 12 m tall and 30 cm in diameter with an umbrella-shaped open crown (unless repeatedly coppiced). It is a softwood tree with timber of low quality. The bark is corky and gummy. Leaves are alternate, oddly bi- or tri-pinnate compound, triangular in outline, and 20–70 cm long. All parts of the *Moringa* tree are edible but the roots, which are used as a condiment in the same way as horseradish, contain the alkaloid spirochin, a potentially fatal nerve paralyzing agent. *Moringa* species have long been recognized by folk medicine practitioners as having value in tumor therapy. Recently, the phytochemicals were shown to be potent inhibitors of phorbol ester (TPA)-induced Epstein–Barr virus early antigen activation in lymphoblastoid (Burkitt's lymphoma) cells [15, 16].

The major objective of this study was to evaluate the synergistic interaction of active phytocompounds of leaf extract with silver nitrate solution for the formation of AgNPs and also to evaluate the catalytic property of AgNPs synthesized from *M. oleifera* on the reduction of methylene blue (MB) dye.

6.2 Materials and Methods

6.2.1 Collection of Leaves

Healthy plant leaves of *Moringa oleifera* Lam was collected periodically during 2016. Botanical examination was performed in the Laboratory of Botany, YVU University, Kadapa. The plant material was washed repeatedly with running tap water thrice and then rinsed with distilled water, dried under low sunlight for 7 days until the plant parts become well dried for grinding. After drying, the plant material was first pulverized and was grinded to a fine powder in a grinder and kept in a well-closed container in a dry place.

6.2.2 Preparation of Aqueous Leaf Extracts of *Moringa oleifera*

The dried leaf powder was used for the preparation of aqueous leaf extract of *Moringa oleifera*; 10 g of fine dried leaf powder was weighed and boiled in 100 ml of distilled water at 90°C for 15 minutes, and then the extract was cooled to room temperature. This extract was separated from the residues by filtering through Whatman No. 1 filter paper and used for further experiments.

6.2.3 Phytochemical Screening

Prepared extract was subjected to preliminary phytochemical screening in order to qualitatively determine some of the secondary metabolites: phenols, carbohydrates, proteins, tannins, alkaloids, flavonoids, steroids, glycosides, and diterpenes using appropriate methods [17, 18].

6.2.4 Preparation of 8 mM Solution AgNO$_3$

An accurately weighed 0.135 g of silver nitrate was dissolved in 100 ml double distilled water and stored in amber color bottle until further use.

6.2.5 Synthesis of AgNPs

For the synthesis of AgNPs, 5 ml of *M. oleifera* leaf broth was taken in a conical flask separately and placed on a magnetic stirrer with hot plate. To this, 45 ml of 8 mM $AgNO_3$ solution was added dropwise with constant stirring, 440 rpm at 50–60°C. The experiment was performed in a 250 ml Erlenmeyer flask. The color change of the solution was checked periodically. The noted change from straw yellow to reddish brown color suggested the formation of AgNPs. Synthesis of AgNPs was monitored by visual observation and UV-visible spectra analysis.

6.2.5.1 Separation of AgNPs

The synthesized AgNPs were separated by centrifugation using a REMI centrifuge at 8000 rpm for 15 min. The process was carried out thrice to get rid of any uncoordinated biomolecules. The supernatant liquid was re suspended in the sterile double distilled water. After the desired reaction period, the supernatant liquid was discarded, and the pellets were collected. The obtained pellet solution was transferred to a sterile petridish and incubated in a hot air oven at 60°C overnight and stored at 4°C for further use.

6.2.5.2 Metal–plant interaction with color formation

The aqueous extract was mixed with the prepared silver nitrate solution and incubated at room temperature. During incubation, metal ions present in the solution interact with the plant phytoconstituents and gets converted from pale yellow to dark brown color. The intensity of the color increased with time. The time duration for color change is primarily due to the excitation of surface plasmon vibrations in AgNPs.

6.2.5.3 UV-visible spectra analysis of the nanoparticles

The reduction of pure silver ions was observed by measuring the UV-visible spectrum of the reaction mixture at different regular time intervals by taking 1 ml of distilled water as blank and 1 ml of reaction mixture each for the analysis. The spectral analysis was done using an UV-visible spectrophotometer evolution 201 series at a resolution of 1 nm from 300 to 800 nm.

6.2.6 Preparation of MB Dye Solution

Typically 1 mg of MB dye was added to 100 ml of double distilled water used as stock solution. The catalytic activity of synthesized AgNPs was carried out in a 3.5 ml capacity quartz cuvette, and absorbance values were monitored using UV-visible spectrophotometer.

6.2.6.1 Evaluation of the effect of synthesized AgNPs on the reduction of MB by *M. oleifera* leaf extract

In order to assess the catalytic activity of synthesized AgNPs, two reactions were carried out as described by Edison and Sethuraman. In the first reaction, 1 ml of MB dye was mixed with 0.2 ml of aqueous leaf extract of *M. oleifera* and 1.8 ml of water, and the reaction was monitored. In the second reaction, 1 ml of MB was mixed with 0.2 ml of aqueous leaf extract of *M. oleifera* and 1.8 ml of synthesized AgNPs [100 mg/ml], and the reactions were monitored after 30 min. The values of absorption maxima of the reaction mixture were calculated using UV-visible spectrophotometer and compared with that of MB. Concentration of the dye during degradation was calculated using the absorbance value at 664 nm.

Percentage of dye degradation was estimated by the following formula:

$$\% \text{ Decolorization} = 100 \times \frac{(C_0 - C)}{C_0}$$

where C_0 is the initial concentration of dye solution and C is the concentration of dye solution after photocatalytic degradation.

6.3 Results and Discussion

6.3.1 Phytochemical Screening

Phytochemicals in medicinal plants have been reported to be the active principles responsible for the pharmacological potentials of plants. Phytochemical screening of *M. oleifera* showed the presence of phenols, carbohydrates, proteins, tannins, alkaloids, flavonoids, steroids, glycosides and diterpenes (Table 6.1). The bioactivity of plant extracts is attributed to phytochemical constituents. For instance, plants rich in tannins have antibacterial potential due to character that allows them to react with proteins to form stable water-soluble compounds, thereby killing the bacteria by directly damaging its cell membrane. Flavonoids are a major group of phenolic compounds reported for their antiviral and antibacterial properties. Phenolic compounds possess hydroxyl and carboxyl groups, and plants with high content of phenolic compounds are one of the best candidates for nanoparticles synthesis.

Table 6.1   Qualitative phytochemical screening of aqueous leaves extract of *M. oleifera*

S. No.	Chemical Constituents	Observation
1	Alkaloids	+ve
2	Glycosides	+ve
3	Saponins	−ve
4	Steroids and terpenoids	+ve
5	Carbohydrates	+ve
6	Flavonoids	+ve
7	Proteins	+ve
8	Amino acids	+ve

'+' = present; '−' = absent

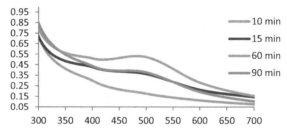

Figure 6.1   Visual identification of silver nanoparticles synthesized by *M. oleifera* leaf extract.

6.3.2 UV-Vis Analysis

The color change of the reaction mixture from light yellow to dark brown was observed within 60 minutes, which indicated the formation of AgNPs as seen in Figure 6.1. This change is due to the excitation of free electrons in nanoparticles which gives the surface plasmon resonance absorption band by the combined vibration of electrons of the metal nanoparticles in resonance with light wave. Metal nanoparticles display different colors in solution due to their optical properties.

In this study, the color of the freshly prepared aqueous extract obtained from the leaf of *M. oleifera* changed when silver nitrate solution was added. The reduction of pure Ag^+ ions was monitored by measuring the UV-visible spectrum of the reaction medium at regular intervals of 10, 15, 60, and 90 min (Figure 6.2). The appearance of reddish brown color within 60 minutes indicates the formation of AgNPs. A stable absorption peak throughout the experiment ranges between 440 and 490 nm. After 2 hours, no significant color change was observed, and increased concentrations of silver nitrate resulted in a brown solution of nano silver indicating the completion of the reaction.

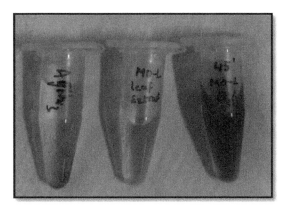

Figure 6.2 UV-Vis absorption spectra of aqueous silver nitrate with *M. oleifera* leaf extract at different time intervals.

6.3.3 Catalytic Activity of Synthesized AgNPs on Reduction of MB by *M. oleifera* Leaf Extract

6.3.3.1 Visual observation

It is a well-known fact that AgNPs and their composites show greater catalytic activity in the area of dye reduction and remove the reduction of MB by arsine in the presence of AgNPs, and this study aims to prove the reduction of MB by natural green aqueous extract of *M. oleifera* containing AgNPs.

Pure MB dye has a (λ_{max}) value of 664 nm. Degradation of dye was visually observed by the change in color from deep blue to light blue. Finally, the degradation process was completed at the end of 5th hour and was identified by the change of reaction mixture from color to colorless (Figure 6.3). The control exhibited no color change. This reveals AgNPs can act as an electron transfer mediator between the extract and MB by acting as a redox catalyst, which is often termed as "electron relay effect." Catalytic activity of AgNPs on the reduction of MB by *Terminalia chebula* fruit extract was reported by Edison and Sethuraman.

6.3.3.2 UV-visible spectrophotometer

Photocatalytic activity of AgNPs on degradation of dye was demonstrated using the MB dye. The degradation of MB was carried out in the presence of AgNPs at different time intervals in the visible region. The absorption spectrum showed the decreased peaks for MB at different time intervals. The completion of the photocatalytic degradation of the dyes is known from the gradual decrease of the absorbance value of dye approaching the base

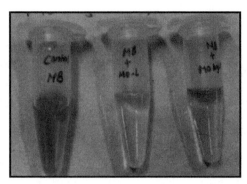

Figure 6.3   Effect of synthesized AgNPs on the reduction of MB dye by *M. oleifera* leaf extract.

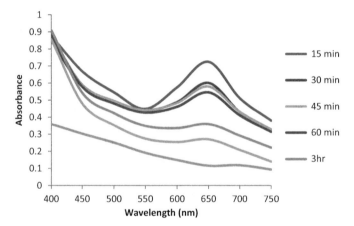

Figure 6.4   UV-visible spectrum of MB dye reduction by MO-L in the presence of AgNPs.

line and increased peak for AgNPs. While decreasing the concentration of dye, UV spectra show typical SPR band for AgNPs at 5 hours of exposure time (Figure 6.4). The percentage of degradation efficiency of AgNPs was calculated as 73.11% at 5th hour. The degradation percentage increased because of the increase in the exposure time of dye AgNPs. Absorption peak for MB dye was centered at 664 nm in the visible region which diminished, and finally, it disappeared while increasing the reaction time, which indicates that the dye had been degraded.

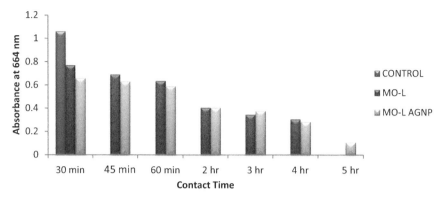

Figure 6.5 Effect of contact time on the adsorption of MB dye at 664 nm by MO-L in the presence of AgNPs.

6.3.3.3 Effect of contact time on adsorption of MB dye

The adsorption of MB dye onto AgNPs was studied as a function of contact time (Figure 6.5) which showed that the dye removal percentage increased on increasing the contact time and reached equilibrium after 5 hours. Karimi et al. reported the adsorption of methyl orange using AgNPs loaded on activated carbon. They observed that the contact time for methyl orange solutions to reach equilibrium was 15 and 18 min for 10 and 20 mg/L of concentration, respectively.

6.4 Conclusion

In this study, AgNPs were green synthesized using the extract of *M. oleifera* leaves. Water-soluble organic compounds present in the plant material were mainly responsible for reducing Ag ions to nano-sized Ag particles. The characterization was done by visual examination using UV-visible spectroscopy. The catalytic activity of green synthesized AgNPs was evaluated by choosing MB dye. Therefore, it can be concluded that the AgNPs synthesized using the leaf extract of *M. oleifera* exhibit good catalytic activity in the dye reduction process. Further research is required to evaluate the mechanism of dye reduction by aqueous leaf extract of *M. oleifera* using AgNPs as nanocatalyst.

Acknowledgement

SAM is thankful to Maulana Azad National Fellowship (MANF), New Delhi, for providing the financial assistance in the form of MANF-JRF to carry out the research work.

References

[1] X. Li, H. Xu, Z. Chen and G. Chen. Biosynthesis of nanoparticles by microorganisms and their applications. Journal of Nanomaterials, vol. 2011, pp. 16, Article ID 270974, 2011.

[2] O. V. Salata. Applications of nanoparticles in biology and medicine. Journal of Nanobiotechnology, vol. 2, article 3, 2004.

[3] S. Kokura, O. Handa, T. Takagi, T. Ishikawa, Y. Naito and T. Yoshikawa. Silver nanoparticles as a safe preservative for use in cosmetics. Nanomedicine, vol. 6, no. 4, pp. 570–574, 2010.

[4] O. Højberg, N. Canibe, H. D. Poulsen, M. S. Hedemann and B. B. Jensen. Influence of dietary zinc oxide and copper sulfate on the gastrointestinal ecosystem in newly weaned piglets. Applied and Environmental Microbiology, vol. 71, no. 5, pp. 2267–2277, 2005.

[5] D. Roe, B. Karandikar, N. Bonn-Savage, B. Gibbins and J. B. Roullet. Antimicrobial surface functionalization of plastic catheters by silver nanoparticles. Journal of Antimicrobial Chemotherapy, vol. 61, no. 4, pp. 869–876, 2008.

[6] E. J. Ferńandez, J. García-Barrasa, A. Laguna, J. M. López-De-Luzuriaga, M. Monge and C. Torres. The preparation of highly active antimicrobial silver nanoparticles by an organometallic approach. Nanotechnology, vol. 19, no. 18, Article ID 185602, 2008.

[7] O. Choi, K. K. Deng, N. -J. Kim, L. Ross Jr., R. Y. Surampalli and Z. Hu. The inhibitory effects of silver nanoparticles, silver 8 Bioinorganic Chemistry and Applications ions, and silver chloride colloids on microbial growth. Water Research, vol. 42, no. 12, pp. 3066–3074, 2008.

[8] S. Gurunathan, K. Kalishwaralal, R. Vaidyanathan, D. Venkataraman, S. R. K. Pandian and J. Muniyandi, et al. Biosynthesis, purification characterization of silver nanoparticles using Escherichia coli. Colloids and Surfaces B: and Biointerfaces, vol. 74, pp. 328–335, 2009.

[9] Y. Y. Chen, C. A. Wang, H. Y. Liu, J. S. Qiu and X. H. Bao. Ag/SiO2: A novel catalyst with high activity and selectivity for hydrogenation

of chloronitrobenzenes. Chemical Communications, vol. 42, pp. 5298–5300, 2005.

[10] P. Setua, A. Chakraborty, D. Seth, M. U. Bhatta, P. V. Satyam and N. Sarkar. Synthesis, optical properties and surface enhanced Raman scattering of silver nanoparticles in nonaqueous methanol reverse micelles. Journal of Physical Chemistry C, vol. 111, pp. 3901–3907, 2007.

[11] R. M. Slawson, J. T. Trevors and H. Lee. Silver accumulation and resistance in *Pseudomonas stutzeri*. Archives Microbiology, vol. 158, pp. 398–404, 1992.

[12] G. J. Zhao and S. E. Stevens. Multiple parameters for the comprehensive evaluation of the susceptibility of *Escherichia coli* to the silver ion. Biometals, vol. 11, pp. 27–32, 1998.

[13] U. P. Parashar, S. S. Preeti and A. Srivastava. Bio inspired synthesis of silver nanoparticles. Digest Journal of Nanomaterials and Biostructures, vol. 4(1), pp. 159–166, 2009.

[14] M. Bala and V. Arya. Biological synthesis of silver nanoparticles from aqueous extract of endophytic fungus *Aspergillus fumigatus* and its antibacterial action. International Journal of Nanomaterials and Biostructures, vol. 3(2), pp. 37–41, 2013.

[15] A. P. Guevara, C. Vargas, H. Sakurai, Y. Fujiwara, K. Hashimoto, T. Maoka, M. Kozuka, Y. Ito, H. Tokuda and H. Nishino. An antitumor promoter from *Moringa oleifera* Lam. Mutation Research, vol. 440, pp. 181–188, 1999.

[16] A. Murakami, Y. Kitazono, S. Jiwajinda, K. Koshimizu and H. Ohigashi. Niaziminin, a thiocarbamate from the leaves of *Moringa oleifera*, holds a strict structural requirement for inhibition of tumor-promoter induced Epstein-Barr virus activation. Planta Medica, vol. 64, pp. 319–323, 1998.

[17] D. Rangari. Pharmacognosy and phytochemistry, 1st Edition, vol. 1. Career Publications, Nashik, p. 100, 2002.

[18] K. R. Khanderwal. Practical Pharmacognosy, 19th Edition. Nirali Prakashan, Pune, pp. 149–156, 2008.

7

Comparative Estimation of Nitrogen and Phosphorus Content of Selected Vegetables: Spectrophotometry as a Tool

Arpita Das

Department of Nutrition, Dr. Bhupendra Nath Dutta Smriti Mahavidyalaya, Burdwan 713407, West Bengal, India
E-mail: arpitadas_0202@yahoo.com

7.1 Introduction

Most of the human body mass is made up of six elements: oxygen, carbon, hydrogen, nitrogen, calcium, and phosphorus, among which nitrogen (N_2) and phosphorus (P) constitute approximately 3% and 1% of the human body weight, respectively [1]. Nitrogen and phosphorus are the two vital elements that play an important role in human body, yet too much of them may be harmful. *Nitrogen* is the largest single component of the earth's atmosphere. Nitrogen is present in all living tissues as proteins, nucleic acids, and other molecules. It is a large component of animal waste (e.g., Guano), usually in the form of urea, uric acid, and compounds of nitrogenous products [2]. *Phosphorus* is present in the human body largely in the form of phosphate (PO_4). Up to 90% of phosphorus in the body is found within calcium phosphate crystals in the bones and teeth [3]. Most of the Indian population is vegetarian and vegetables contain appreciable amount of N_2 and P. The composition of vegetables can vary depending upon the cultivar and origin [4]. On this basis, in this study the N_2 and P contents were compared between some common vegetables grown in the landfill sites of Kolkata (Dhapa) and farming field of Sonarpur, India. Dhapa is a locality on the eastern fringes of Kolkata, India. The area consists of landfill sites where the solid wastes of the city of Kolkata are dumped. "Garbage farming" is encouraged in the landfill sites. More than 40%

of green leafy vegetables in the Kolkata markets come from these lands (http://en.m.wikipedia.org/wiki/Dhapa,_India).

The following vegetables were collectedfrom the above-mentioned places as samples for this study:

- Spinach (*Spinacia oleracea*)
- Brinjal (*Solanum melongena*)
- Cauliflower (*Brassica oleracea*)
- Mustard leaves (*Brassica campestris*)
- *Water* (used for cultivation)

The samples collected from Sonarpur and Kolkata Dhapa were considered as Fresh and Dhapa samples, respectively.

7.2 Aims and Objectives

The study aimed to estimate and compare the nitrogen ($NO_{3-}-N$ and NH_3-N) and phosphorus ($PO_4^{3-}-P$) contents in spinach, brinjal, cauliflower, mustard leaves, and water (used in cultivation) collected from Sonarpur and Kolkata Dhapa.

7.3 Materials and Methods

The following methods were used:

(1) Spectrophotometric measurement (at 220 nm) of nitrogen ($NO_{3-}-N$) in water samples (using 1 N of HCl).

(2) Spectrophotometric measurement (at 410 nm) of nitrogen (NH_3-N) in plant materials using Nessler's method.

(3) Spectrophotometric measurement (at 690 nm) of phosphorus ($PO_4^{3-}-P$) in water samples and plant materials by stannous chloride reduction method [5].

All plant samples were first dried, powdered, and digested with acid and diluted with water. Standard curves were constructed for each method, and then the light transmittance of the samples was measured under spectrophotometer after mixing with reagents. Using sample absorbance, sample concentrations were directly obtained from the standard curve.

7.4 Results

7.4.1 Comparative Study of Fresh and Dhapa Samples

All the fresh and dhapa vegetables as well as respective water samples were compared on their nitrogen and phosphorus content. The nitrogen content of all the vegetables and water were estimated in form of ammonia and nitrate ions respectively. The phosphorus content of vegetables and water were estimated in form of phosphate ion. The following bar diagrams in Figure 7.1 shows the comparative results.

$$\text{Nitrogen } (NH_3-N) \qquad \text{Phosphorus } (PO_4^{3-}-P)$$

7.5 Discussion

According to the results, the nitrogen content of spinach, brinjal, and cauliflower is almost equal in both fresh and Dhapa samples. But Dhapa mustard leaves and Dhapa water sample contain much higher amounts of nitrogen as compared to their respective fresh samples.

The phosphorus content of Dhapa spinach, mustard leaves, and water is slightly higher than their respective fresh samples. Also, Dhapa brinjal and cauliflower have much higher amounts of phosphorus as compared to their respective fresh samples.

7.6 Summary and Conclusion

Nutrients are the constituents in food that must be supplied to the body in suitable amounts. Too much of them may be harmful. Vegetables supply many nutrients besides providing variety to the diet. Vegetables are produced by plants from the nutrients available in the soil, water, and the carbon dioxide from air. Attempts have been made in this study to estimate and compare the nitrogen and phosphorus contents in some common vegetables of two differentlands.The study shows that in most of the cases, the selected vegetables grown in Dhapa contain higher amounts of phosphorus,but the nitrogen content is not significantly higher than that of the respective vegetables of Sonarpur.

[a] Nitrogen (NH₃-N)   Phosphorus(PO³₄⁻-P)

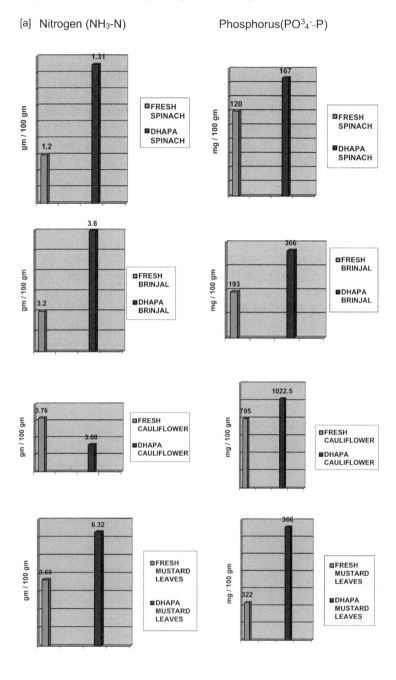

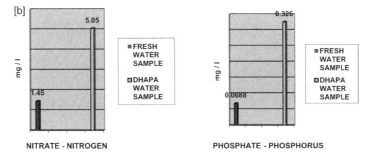

Figure 7.1 Comparative nitrogen (left side) and phosphorus (right side) content in form of [a] ammonia and phosphate ions in spinach, brinjal, cauliflower, mustered leaves and in form of [b] nitrate and phosphate ions in water of fresh and dhapa samples from top to bottom.

The outcome of this study may be due to the following:

(1) Geographical distribution
(2) Soil composition
(3) Climate condition
(4) Maturity of the samples.

References

[1] Harper, H. A., Rodwell, V. W. and Mayes, P. A. (1977). Review of physiological chemistry, 16th ed. Los Altos, California: Lange Medical Publications.

[2] Lide, D. R. (c1977). CRC handbook of chemistry and physics. Cleveland, Ohio : CRC Press, pp. 1913–1995.

[3] Srilakshmi, B. (2003). Food science. New Delhi: New Age International (P) Limited Publishers.

[4] Belitz, H.-D., Grosch, W. and Schieberle, P. (2009). Food chemistry, 4th Revised & Extended Edition. Springer, p. 770, Berlin, Germany.

[5] [a] APHA (American Public Health Association). (1998). Standard method for examination of water and waste water, 20th edition, Washington, DC, USA, pp. 4.115–4.118 (4500- APHA NO_3 B&E). [b] APHA (American Public Health Association). (1998). Standard method for examination of water and waste water, 20th edition, Washington, DC, USA, pp. 4.145–4.146 (4500-P D).

8

UV as a Potential Signal for Communication in Butterflies and Moths

Monalisa Mishra

Department of Life Science, NIT Rourkela, Rourkela, Odisha 769008, India
E-mail: mishramo@nitrkl.ac.in; monalisamishra2010@gmail.com

8.1 Introduction

Ultraviolet (UV) signal is found in the wings of many butterflies and moths. UV reflectance arises due to the arrangement of scales which causes interference. Only a specific portion of the wing possesses UV signal. Mostly, they include the white or bright spots or patterns present in the wing. The UV signal present in the wing of butterflies and moths is to provide intra- as well as interspecific signal communication [1]. Eyespot or eyelike spot, which is used to protect the animal from the predator, also possesses UV signal [2]. The butterflies which possess UV signal in the eyespot have better protecting ability than the one without it [3]. Each species has its own UV reflection pattern to reduce the confusion with the nearest species. Although we are not able to see the UV signal since human eye has sensitivity only in the visible range, butterflies and moths use it as the main signal for communication [4]. This study analyzes the UV reflection pattern of two different families of Lepidoptera (Papilionidae and Nymphalidaea) not described in earlier studies.

8.2 Materials and Methods

Butterflies from Papilionidae and Nymphalidae were collected from NIT Rourkela, Rourkela, Odisha, India, 22.25 N, 84.90 E during summer 2017. Butterfly species having more than 6 numbers were used for our analysis; 4 members from Papilionidae and 5 members from Nymphalidae were used for this study. At least 6 butterflies from each species were analyzed. Intact butterflies were imaged with digital camera, under epiwhite light and UV light, and transferred to Image J for further processing.

8.3 Results

8.3.1 Analysis of Papilionidae Family

UV signal of four different members of Papilionidae such as *Papilio demoleus* (Common lime), *Graphium doson* (Common jay), *Papilio polytes* stichius form and *Papilio polytes* cyrus form were analyzed under both epiwhite and UV light. *Papilio demoleus* possesses yellowish white patches throughout the wing and an oval-shaped patch with a combination of orange, blue, and black color (Figure 8.1A). Although the oval patch is visible under normal light, it is not seen under epiwhite and UV light (Figure 8.1A′, A″). All the yellowish white patches possess UV signals. *Graphium doson* possesses many bluish white patches and red and black rings in the wing (Figure 8.1B). The red and black patches are not seen under epiwhite and UV

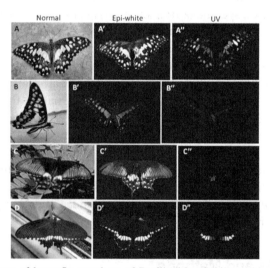

Figure 8.1   Images of butterfly members of Papilionidae family under normal, epiwhite, and UV light. *Papilio demoleus* (A) under normal camera, (A′) under epiwhite, and (A″) under UV light. *Graphium doson* (B) under normal camera, (B′) under epiwhite, and (B″) under UV light. *Papilio polytes* stichius form (C) under normal camera, (C′) under epiwhite, and (C″) under UV light. *Papilio polytes* cyrus form (D) under normal camera, (D′) epiwhite, and (D″) under UV light.

light (Figure 8.1B′, B″). However, all the bluish white patches possess UV signals. *Papilio demoleus* stichius form possesses more orange patches in the hindwing (Figure 8.1C). Although orange patches show some signal under epiwhite (Figure 8.1C′), the signals are completely disappeared under UV light (Figure 8.1C″). UV linings are found in the forewing closer to the body. In cyrus form, the white patches are found in both the forewing and hindwing regions (Figure 8.1C"). The forewing possesses UV in the marginal region, whereas the hindwing possesses UV signals at the central as well as marginal region. Within the same species, it is quite striking to see the variation of UV signal in both forms of *P. demoleus.*

8.3.2 Analysis of Nymphalidae Family

Five different members of this family were analyzed under epiwhite and UV light (Figure 8.2). Female *Hypolimnus bolina* is a brownish black colored butterfly with white patches in the margin. The butterfly was checked for the coloration pattern under epiwhite and UV light. The female possesses white pattern in both the forewing and hindwing. The white pattern is more intense and beautiful in the hindwing (Figure 8.2A). Although the white pattern is still visible under epiwhite (Figure 8.2A′), the wing signal becomes more prominent under UV light (Figure 8.2A″). *Junonia almana,* another member of this family, possesses orange color body with two eyespots on the forewing and hindwing (Figure 8.2B). The eyespot of hindwing is larger than the forewing, is oval shaped, and possesses black and orange colored concentric rings (Figure 8.2B′). The forewings possess many black and white circular shaped concentric rings. Only the forewing eyespot possesses UV signal at the center (Figure 8.2B″). *Danaus chrysippus* is an orange colored butterfly with white lining found in both the forewing and hindwing with many white patterns in the forewing (Figure 8.2C). Although a very bright pattern is visible under epiwhite (Figure 8.2C′), only white patches are known to possess UV signal (Figure 8.2C″). *Junonia atlites,* a pale lavender brown color butterfly, possesses many eyespots throughout the margin. The eyespot contains white orange and black patches of unequal size (Figure 8.2D). The pattern is prominent under epiwhite (Figure 8.2D′). Throughout the body, mild UV signal is found except the dark patches (Figure 8.2D″). *Melanitis leda* exhibits polyphenism in its wings (Figure 8.2E, E′). The dark brown coloration is found in the winter (Figure 8.2E), whereas the pale brown coloration is observed in the rainy season (Figure 8.2E′). Eyespots are found in the dorsal forewing and hindwing of the *M. leda.* The eyespots

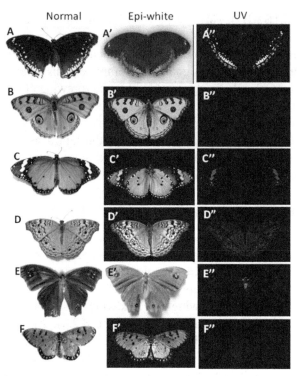

Figure 8.2   Images of butterflies of Nymphalidae family under normal camera, epiwhite, and UV light. *Hypolimnas bolina* (A) under normal camera, (A′) under epiwhite and (A″) under UV light. *Junonia almana* (B) under normal camera, (B′) under epiwhite (B″), and under UV light. *Danaus chrysippus* (C) under normal camera, (C′) under epiwhite, and (C″) under UV light. *Junonia atlites* (D) under normal camera, (D′) under epiwhite, and (D″) under UV light. *Melanitis leda* (E) under normal camera, (E′) under epiwhite, and (E″) under UV light. *Acraea terpsicore* (F) under normal camera, (F′) under epiwhite, and (F″), under UV light.

possess white spot at the center. The white spot possesses UV signal in it (Figure 8.2E″). *Acraea terpsicore* is an orange colored butterfly with black lining in the margin. The hindwing possesses thick black lining with white oval-shaped spots (Figure 8.2F). This pattern is visible under epiwhite (Figure 8.2F′). Under UV, the white spots that possess UV signal are only visible (Figure 8.2F″).

8.4 Discussion

All the examined species of the two different families Papilionidae and Nymphalidae possess unique UV patterns. This study deciphers the variation in the UV pattern in two different forms of *Papilio polytes*. In all the examined Papilionidae species, the bright white patches possess UV signals. The orange colored patches do not possess UV signals. However, insects are known to possess infrared signals in the eye. Does this orange patch contain infrared signal in it? This is an open question at this moment and needs further investigation.

Hypolimnus bolina male is well known for its oval UV patches surrounded by iridescent blue scales in its wings [9]. Females prefer the male based on the overstated UV patch. This study reports multiple bright UV patterns in the female wings as well. Like *H. bolina*, *Colias eurytheme* male also possesses iridescent UV patches in its wings. Building of iridescent structure must be expensive in terms of nutrition, and thus, many of the members instead of making iridescent patches prefer to have UV as a signal for communication [8]. The UV reflectance is also produced by the fine nanostructural arrangement of the lamellae within the scales. Presence of pterin like pigments further enhances the UV signal within the wing [10]. Male signaling helps female to provide time for egg laying and feeding [6, 7]. *Junonia almana* possesses UV signal only in the center part of the forewing. *Danaus chrysippus* possesses UV signal in the white spots present in the forewing and hindwing. *Junonia atlites* possesses a lot of eyespots in the margin of its wings to protect itself from birds attack [11], and there was no UV signal in the eyespot. In *Melanitis leda*, only the center part of the eyespot possesses UV signal [5]. *Acraea terpsicore* possesses UV signal in the margin of the wings probably to protect itself from birds attack.

8.5 Conclusion

All the bright spots and patches present in the wings of Papilonidae and Nymphalidae possess UV signals which are probably for signal communication.

References

[1] Osorio, D. and Vorobyev, M., 2008. A review of the evolution of animal colour vision and visual communication signals. *Vision Research*, 48(20), pp. 2042–2051.

[2] Olofsson, M., Vallin, A., Jakobsson, S. and Wiklund, C., 2010. Marginal eyespots on butterfly wings deflect bird attacks under low light intensities with UV wavelengths. *PLoS ONE*, 5(5), p. e10798.

[3] Silberglied, R.E. and Taylor, O.R., 1973. Ultraviolet differences between the sulphur butterflies, Colias eurytheme and C. philodice, and a possible isolating mechanism. *Nature*, 241(5389), p. 406.

[4] Rutowski, R.L., Macedonia, J.M., Merry, J.W., Morehouse, N.I., Yturralde, K., Taylor-Taft, L., Gaalema, D., Kemp, D.J. and Papke, R.S., 2007. Iridescent ultraviolet signal in the orange sulphur butterfly (Colias eurytheme): spatial, temporal and spectral properties. *Biological Journal of the Linnean Society*, 90(2), pp. 349–364.

[5] Eram, S., Sabat, D., Sahu, B.B. and Mishra, M., 2016. Structural Variations in Wing Patterning of Seasonal Polyphenic Melanitis leda (Satyrinae). *Microscopy Research*, (04) No. 0471988.

[6] Graham, S.M., Watt, W.B. and Gall, L.F., 1980. Metabolic resource allocation vs. mating attractiveness: adaptive pressures on the "alba" poly morphism of Colias butterflies. *Proc Natl Acad Sci USA*, 77:3615–3619.

[7] Gilchrist, G.W. and Rutowski, R.L., 1986. Adaptive and incidental con sequences of the alba polymorphism in an agricultural population of Colias butterflies: female size, fecundity, and differential dispersion. *Oecologia*, 68:235–240.

[8] Kemp, D.J. and Rutowski, R.L., 2007. Condition dependence, quantitative genetics, and the potential signal content of iridescent ultraviolet butterfly coloration. *Evolution*, 61(1):168–83.

[9] Kemp, D.J., 2007. Female butterflies prefer males bearing bright irides-cent ornamentation. *Proc Biol Sci*, 274(1613):1043–7.

[10] Rutowski, R.L., Macedonia, J.M., Morehouse, N. and Taylor-Taft, L., 2005. Pterin pigments amplify iridescent ultraviolet signal in males of the orange sulphur butterfly, Colias eurytheme. *Proceedings of the Royal Society B: Biological Sciences*, 272(1578), pp. 2329–2335.

[11] Olofsson, M., Vallin, A., Jakobsson, S. and Wiklund, C., 2010. Marginal eyespots on butterfly wings deflect bird attacks under low light intensities with UV wavelengths. *PLoS One*, 5(5), p. e10798.

9

−NH Protons Containing Heterocycles: Colorimetric Chemosensor for Fluoride Ion

Tandrima Chaudhuri

Department of Chemistry, Dr. Bhupendranath Dutta Smriti Mahavidyalaya, Burdwan, West Bengal, India
E-mail: tanchem_bu@yahoo.co.in

Comparatively less fluorescent imidazoles containing −NH ring can efficiently sense F^- ion in a polar medium like acetonitrile. Photophysics can establish the interaction of tetra-n-butylammonium fluoride with two different benzimidazoles via stable isosbestic and isoemissive formation. Further, proton nuclear magnetic resonance hydrogen-1 shows the interaction as abstraction of −NH proton by the F^-.

9.1 Introduction

Molecular level fluoride sensors can broadly be classified into four types: (1) molecules capable of sensing anions through hydrogen bonding interactions containing N–H, C–H, and O–H groups [1], (2) anion–π interactions [2], (3) Lewis acid-base interactions [3], and (4) anion induced chemical reactions [4]. This study is particularly projected towards H-bonding interaction. Especially at lower F^- concentrations, the H-bonding interactions between the N–H/O–H fragment of the receptor and the F^- ion are observed; however, by means of Brønsted acid-base interactions an excess of F^- ion leads to deprotonation [5].

With this evidence, two different benzimidazoles (**R-1** and **R-2**) containing −NH proton (shown in Figure 9.1) to interact with tetra-n-butylammonium fluoride (TBAF) in a polar organic medium (CH$_3$CN) were chosen. Visible color change was observed in all the cases (Figure 9.2).

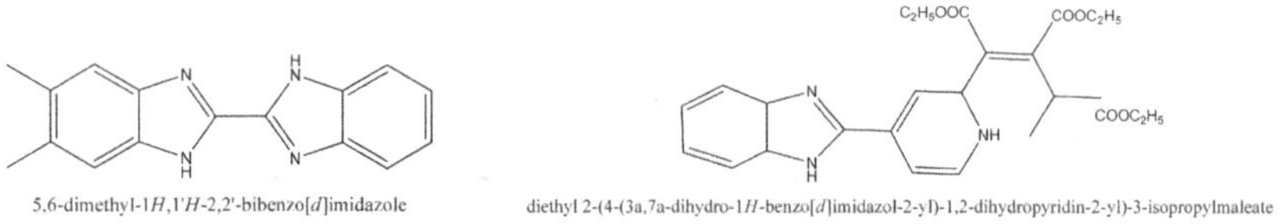

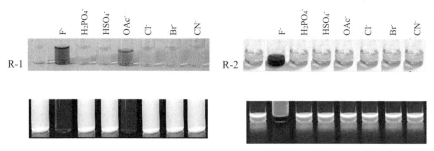

5,6-dimethyl-1*H*,1'*H*-2,2'-bibenzo[*d*]imidazole

diethyl 2-(4-(3a,7a-dihydro-1*H*-benzo[*d*]imidazol-2-yl)-1,2-dihydropyridin-2-yl)-3-isopropylmaleate

Figure 9.1 Structure of the benzimidazoles **R-1** and **R-2**.

Figure 9.2 Color change of **R-1** (2 mM) and **R-2** (2 mM) on addition of the TBA$^+$ salts (5 mM).

The photophysical study and proton nuclear magnetic resonance hydrogen-1 (^{1}H NMR) measurements were reported that verified the mode of interaction of the receptors with the TBAF and other anions.

9.2 Materials and Methods

Acetonitrile used as a solvent was of the high performance liquid chromatography grade. All of the tetrabutyl amonium salts were purchased from Aldrich. The concentration of all two receptors **R-1** and **R-2** were made in the range of 10^{-4}–10^{-5} M in all the spectral measurements. The concentrations of all the tetrabutyl ammonium salts were of the order of $\sim 10^{-3}$–10^{-4} M.

9.2.1 Instruments Used

The absorption ultraviolet-visible (UV-Vis) spectral measurements were performed with a Shimadzu UV 1800 spectrophotometer fitted with an electronic temperature controller unit (TCC–240 A). The steady state fluorescence emission and excitation spectra were recorded with a Hitachi F-4500 spectrofluorometer equipped with a temperature controlled cell holder. Temperature was controlled within ± 0.1 K by circulating water from a constant temperature bath (Heto Holten, Denmark). Further, ^{1}H NMR spectra were recorded using Bruker 300 MHz spectrometer at 298 K in CD$_3$CN.

9.3 Results and Discussion

9.3.1 Photophysical Study

It is a well-known fact that benzimidazole contains $-$NH proton within the ring, and is very useful for selective binding of anions. Thus anions having higher basicity like F$^-$, OAc$^-$, H$_2$PO$_4-$, HSO$_4-$, Cl$^-$, Br$^-$, CN$^-$, I^-, etc. can act as analytes for detection by these receptors. Our assessment in ambient conditions showed that upon the addition of various anions, a remarkable color change from pale yellow of **R-1** to deep green only in the presence of F$^-$ and slightly in the presence of AcO$^-$ in solution took place; however, the change of yellow color of the **R-2** to deep brown in the presence of F$^-$ in the solution indicates that receptors **R-1** and **R-2** have the potential to act as specific sensors for the F$^-$ ion (Figure 9.2).

The selective anion recognition properties of the receptors **R-1** and **R-2** towards different anions were studied experimentally by naked eye, UV-visible, fluorescence, and ^{1}H-NMR spectroscopy. Receptors were synthesized in the reported methods [6] and research regarding their sensing application is still in progress.

The receptor **R-2** exhibited four absorption bands at 252 nm, 282 nm, 305 nm, and a very strong band at 366 nm with a hump at 385 nm. After gradual addition of the TBAF solution into the **R-2** solution an ample spectral change was observed (Figure 9.3a). The addition of the F$^-$ salt solution to the **R-2** in acetonitrile showed a blue-shifted absorption band from 252.4 nm to 249.8 nm with the appearance of a prominent hump at 217 nm. However, the 282 nm peak (Figure 9.3a) diminished after addition of the TBAF salt solution, the 305 nm peak was red shifted to 311.6 nm and the longer wavelength dual peak of 366 nm and 385 nm was turned to a prominent single peak at 385 nm, and simultaneously a new band appeared at 400 nm which was the charge transfer band. The red shift in short wavelength and broadening of the longest wavelength peak may be due to hydrogen bond interaction of F$^-$ with **R-2** and yellow color of the **R-2** turning deep brown at the same time (Figure 9.2) may be due to complete deprotonation. Two well defined isosbestic points were observed at 253.40 nm and 372 nm which indicated the formation a single complex between the **R-2** and F$^-$. The stoichiometry

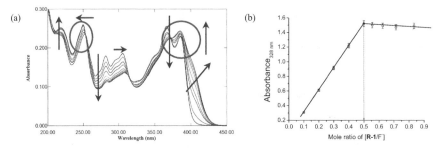

Figure 9.3   (a) Ground state titration of the **R-2** by the TBAF in acetonitrile. Concentrations of the TBAF (mM): 0.00, 0.76, 1.48, 2.16, 2.82, 3.45, 4.05, 4.62, 5.17, 5.70, 6.20, 6.69, 7.16, and 7.61 at a fixed concentration (0.51 μM) of the **R-2** solution in acetonitrile. (b) Ground state Job's plot for the complex of **R-1** and F⁻.

of the (**R-2**:F⁻) complex was proposed to be 1:2 as that of the **R-1** obtained in Job's plot (Figure 9.3b).

Interaction with other variety of anions such as $H_2PO_4^-$, HSO_4^-, Cl^-, Br^-, CN^-, and I^- was also observed but selectively the color of the solution of all the three **R** changed with the addition of F⁻ and of AcO⁻ only (Figure 9.2a). This result is strong evidence that the red shift of the absorption band is probably due to the disruption of the −NH proton.

In good uniformity with the UV-visible study, the **R-2** showed a specific emission response to the F⁻ ion also. The emission properties of the **R-2** in the presence of various ions were examined in CH_3CN. When the **R-2** was excited at 366 nm in a UV cabinet the free receptor showed only significant fluorescence, whereas fluoride added solution of the **R-2** demonstrated striking blue color fluorescence (Figure 9.2). As shown in Figure 9.4a upon progressive addition of fluoride fluorescence intensity of the **R-2** gradually decreased and showed a red shift. A distinct isoemissive point appeared in the case of F⁻ (Figure 9.4) which indicated clear equilibrium interaction also takes place in the excited state. But for other anions negligible fluorescence change was observed. The Stern–Volmer binding constant of the **R-2** with F⁻ is $16.19 \times 10^4 \ M^{-1}$.

The ^1H NMR study also indicated disruption of N–H proton on addition of F⁻ salt in both **R-1** and **R-2** solutions in CD_3CN.

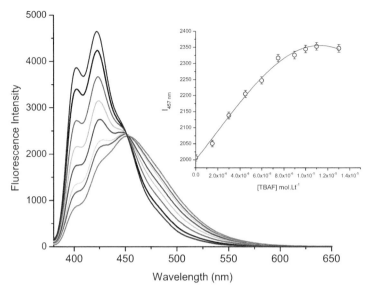

Figure 9.4   Fluorescence titration of the **R-2** by the TBAF in acetonitrile. Concentrations of the TBAF (mM): 0.00, 0.76, 1.48, 2.16, 2.82, 3.45, 4.05, 4.62, 5.17, 5.70, 6.20, 6.69, 7.16, and 7.61 at a fixed concentration (0.51 µM) of the **R-2** solution in acetonitrile. Inset: Fluorescence calibration curve at 457 nm.

9.4 Conclusion

Thus, we conclude that organic heterocycles containing the −NH protons are very much likely to interact with the F⁻ salt and hence can be promising as colorimetric sensors.

References

[1] (a) Swinburne, A. N., Paterson, M. J., Beeby, A. and Steed, J. W. "Fluorescent 'Twist−on' Sensing by Induced−Fit Anion Stabilisation of a Planar Chromophore" Chem. Eur. J. 2010, 16, 2714–2718. (b) Veale, E. B., Tocci, G. M., Pfeffer, F. M., Krugera, P. E. and Gunnlaugsson, T. "Demonstration of bidirectional photoinduced electron transfer (PET) sensing in 4-amino-1,8-naphthalimide based thiourea anion sensors" Org. Biomol. Chem. 2009, 7, 3447–3454.

[2] (a) Guha, S. and Saha, S. "Fluoride Ion Sensing by an anion π Interaction" J. Am. Chem. Soc. 2010, 132, 17674–17677. (b) Guha, S., Goodson, F. S., Corson, L. J. and Saha, S. "Boundaries of Anion/Naphthalenediimide Interactions: From Anion-π Interactions to Anion-Induced Charge-Transfer and Electron-Transfer Phenomena" J. Am. Chem. Soc. 2012, 134, 13679–13691.

[3] Wade, C. R., Broomsgrove, A. E. J., Aldridgeand, S. and Gabbaï, F. P. "Fluoride ion complexation and sensing using organoboron compounds." Chem. Rev. 2010, 110, 3958–3984.

[4] (a) Zhang, J. F., Lim, C. S., Bhuniya, S., Cho, B. R. and Kim, J. S. "A Highly Selective Colorimetric and Ratiometric Two-Photon Fluorescent Probe for Fluoride Ion Detection" Org. Lett. 2011, 13, 1190–1193. (b) Mahoney, K. M., Goswami, P. P. and Winter, A. H. "Self-Immolative Aryl Phthalate Esters" J. Org. Chem. 2013, 78, 702–705. (c) Hu, R., Feng, J., Hu, D., Wang, S., Li, S., Li, Y. and Yang, G. "A rapid aqueous fluoride ion sensor with dual output modes." Angew. Chem., Int. Ed. 2010, 49, 4915–4918.

[5] (a) Qu, Y., Hua, J. and Tian, H. "Colorimetric and Ratiometric Red Fluorescent Chemosensor for Fluoride Ion Based on Diketopyrrolopyrrole" Org. Lett. 2010, 12, 3320–3323. (b) Yang, C., Zheng, M., Li, Y., Zhang, B., Li, J., Bu, L., Liu, W., Sun, M., Zhang, H., Tao, Y., Xue, S. and Yang, W. "N-Monoalkylated 1,4-diketo-3,6-diphenylpyrrolo[3,4-c] pyrroles as effective one- and two-photon fluorescence chemosensors for fluoride anions" J. Mater. Chem. A. 2013, 1, 5172–5178.

[6] (a) Mukhopadhyay, C., Ghosh, S. and Butcher, R. J. "An efficient and versatile synthesis of 2, 2'-(alkanediyl)-bis-1*H*-benzimidazoles employing aqueous fluoroboric acid as catalyst: density functional theory calculations and fluorescence studies" Arkivoc. 2010, ix, 75–96. (b) Bagdi, A. K., Santra, S., Monir, K. and Hajra, A. "Synthesis of imidazo [1,2-*a*] pyridines: a decade update" Chem. Commun. 2015, 51, 1555–1575.

MODULE 3

10

Particle Size Control on Wet Micronization of Sulfur Prills for Fertilizer Applications

Suresh Puthiyaveetil Othayoth*, **Nirmit Kantilal Sanchapara,**
Soheb Husenmiyan Shekh, Amarkumar Bhatt
and Sandeep Jasvantrai Parikh

R&D Division, Gujarat State Fertilizers and Chemicals Limited,
Fertilizer Nagar, Vadodara 391750, Gujarat, India
E-mail: posuresh@gsfcltd.com
*Corresponding Author

Application of micronized elemental sulfur as a plant nutrient in agriculture enhances oil content in oil seeds, improves chlorophyll formation, and increases shelf life of some vegetables. Micronization of elemental sulfur with minimum hazards and optimum particle properties is essential for its better application. Colloidal milling with various surfactants at various concentrations is tried with varying milling time to obtain control over particle size. Redispersion of the particles is tested before application to soil. UV-Vis spectrophotometry of fine sulfur particles at varying concentrations in methanol helped estimate the average particle size. This was confirmed using the standard sieve analysis. Field application of the micronized sulfur fertilizer shows a considerable increase in yield when compared with that from the application of commercially available sulfur fertilizer prepared from molten sulfur.

10.1 Introduction

Sulfur is considered as an essential secondary plant nutrient. In recent past, sulfur deposition to soil by its natural sources has considerably reduced, and excessive farming has consumed the available sulfur from the soil, creating a requirement of sulfur supplementation. There have been examples of sulfur supplementation increasing crop production by more than 100% [1]. Plants absorb sulfur as sulfates. Ammonium sulfate is the generally used sulfur fertilizer. The high solubility of sulfate in water causes its loss due to leaching before being absorbed by plants. Insoluble elemental sulfur is an alternative source of sulfur nutrient. Blair [2] compared the retention time of sulfate and elemental sulfur in soils to report ~50% retention of elemental sulfur after 52 weeks against a 100% removal of sulfate. However, the sulfur needs to be solubilized for the plants to absorb. Owing to the bacterial activity in the soil, the elemental sulfur will be oxidized to form sulfates, and it is then subsequently absorbed by plants. Chapman et al. [3] reported the influence of temperature, moisture, and ambient pH on the conversion of S to sulfates. For effective conversion, the sulfur particles should be ~70 μm in size and should be water dispersible while applied in the soil [4].

Preparation of finely divided sulfur in micron size by grinding involves environmental as well as health hazards. Here we discuss a process for the production of micronized sulfur minimizing possibilities of such hazards. Such a process is essential for the production of powdered sulfur with size control for industrial application also. Elemental sulfur used in ammunitions is finely ground to increase burning efficiency. Similarly, sulfur used for rubber processing for manufacturing tyre is ground to very fine powder for increasing process efficiency. This necessitates the estimation of average particle size. Particle size is generally determined using dry or wet sieving techniques [5]. Settling tubes can also be used to estimate the particle size distribution (e.g., [6]). However, these two processes are time-consuming and may hinder the pace of processes in a large-scale production atmosphere. Hence, more sophisticated techniques or instruments may be required.

Light scattering by particles dispersed in a medium can provide information on particle size [7]. Theoretical and experimental studies on using light scattering to estimate particle size have been extensively discussed by Bertero et al. [8]; Wang and Hallett [9]; Tscharnutar [10]; and references therein. Techniques based on laser diffraction and Mie theory and X-ray attenuation are extensively being used in particle size distribution. Our observation on the applicability of UV-absorption spectrophotometer to estimate the average particle size of micronized elemental sulfur is also detailed here.

10.2 Materials and Methods

Pure sulfur prills were obtained from the Gujarat State Fertilizers and Chemicals Limited to produce powder of various size ranges. Purity of the sulfur was established as per Indian Standard IS: 6444 (2012).

A colloidal mill with slotted rotor–slotted stator assembly is used for wet milling sulfur prills. The mill was driven by a 1 HP motor powered by three-phase electricity. The maximum milling volume at a time in the rotor–stator assembly is about 500 ml, which can be further reduced by adjusting the position of the stator. The particle size distribution of various batches was adjusted by varying the milling volume as well as milling time. Demineralized water with suitable dispersing agent was used for wet milling. The colloid thus produced was filtered in a vacuum filtration unit and then dried in an oven at 60°C. Samples of various modal particle size distributions were prepared by varying machine running time. The details of experiments regarding the preparation of samples of various particle size distributions (PSD) are detailed in the subsequent sections.

Standard test sieves of sizes were used to separate the dried sulfur particles into various size groups for UV spectrometer analysis. Accurately weighed amount of the sample from each size group was used to prepare a dispersion of known concentration using methanol as wetting agent. After ultrasonication for 5 minutes, the UV-absorption spectrum of each sample was recorded.

A Shimadzu UV-1800 UV-Vis spectrometer was used in spectrum mode for the reported experiments. The instrument has high wavelength resolution of 1 nm ranging from 190 to 1100 nm. Square cuvettes of 10 mm path length were used for holding the sample. UVProbe software provided by Shimadzu was used for recording the spectra.

The finely divided sulfur was then pressed into tablets after mixing with bentonite to apply to soil as fertilizer.

10.3 Results and Discussion

10.3.1 Preparation of Micronized Elemental Sulfur

The major parameters that can affect the particle size of a material produced in a colloidal mill are run time, choice of dispersant used, and the quantity of dispersant used. Experiments were conducted to establish the optimum conditions for each parameter for producing minimum particle size. During the trials for each parameter, all the other parameters are kept fixed. Figure 10.1 shows the effect of running time of the colloidal mill on particle size. For a given amount of raw material and surfactant quantity, the weight % of the particles below 37 μm increases linearly. A 63% increase in particle mass having size below 37 μm is observed when the milling time is increased from

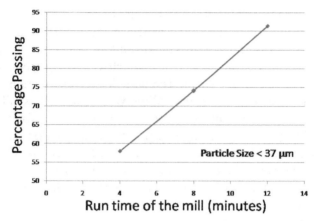

Figure 10.1   Weight % of particles of elemental sulfur when the colloidal mill running time is varied. The amounts of raw materials used in each run remain the same.

4 minutes to 12 minutes. As shown in Figure 10.1, for the current experiment, quantity by weight of the material ground to below 37 μm size is 4.5% of the total material per minute. The implication is that for getting smaller particles, the run time could be increased.

Surfactants help reduce surface tension at the solid–liquid interface. While adsorbed on the surface pores of the particles, the surfactant causes pressure to the grain to help further breakdown. It also contributes to reduce interactions between two grains to resist agglomeration. Hence, the surfactant used while producing the micronized elemental sulfur is expected to have an effect on particle size distribution. Sodium lauryl sulfate (SLS), carboxy methyl cellulose (CMC), and sodium dodecylbenzenesulphonate (SDBS) are used in experiments for a comparison of PSDs (Figure 10.2). When CMC is used, the quantity of coarser particles present in the product is observed to be more than that produced with the use of SLS. Moreover, the use of CMC hinders filtration of the product to remove water. Better results were obtained when SDBS is used. It has been reported that the hydrophobic interaction in a polar environment in the liquid phase may lead to agglomeration of particles [11]. Variation in the particle size with the surfactant used could be due to the difference in the behavior of the polar and non-polar characteristics of the surfactant used.

The quantity of surfactant used seems to have an influence on PSD. As the quantity increases, the percentage weight of particles below 37 μm increases (Figure 10.3). However, the trend, if it can be considered, is towards a

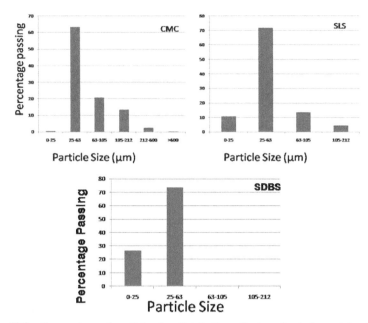

Figure 10.2   Comparison of particle size distribution of micronized elemental sulfur produced with carboxy methyl cellulose (CMC), with sodium lauryl sulfate (SLS) and sodium
dodecylbenzenesulphonate (SDBS). The quantity of surfactant used, the quantity of sulfur
used, and the run time of the mill remain the same. Least fraction of coarser particles is
observed when SDBS is used.

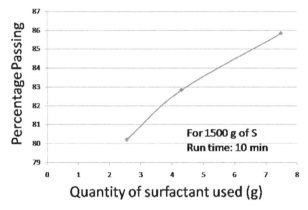

Figure 10.3   Variation in the fraction of micronized elemental sulfur particles having size
below 37 μm when the quantity of surfactant is varied. The run time and quantity of sulfur
used remain the same.

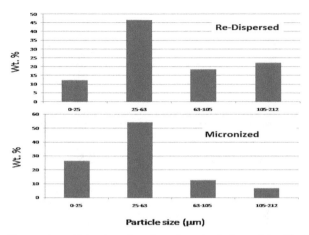

Figure 10.4   Change in particle size when the micronized elemental sulfur is redispersed. Although coarsening is observed, the conglomerates still stay below 212 μm in size.

plateau. For nanoparticles of sulfur, an increase in particle size has been noted with increased concentration of surfactants, reported by Turganbay et al. [11]. However, the surfactant concentrations used by them are considerably lower than what is reported here. It can be proposed that as the particle size decreases, the total surface area increases and the ratio of available surfactant quantity to total surface area decreases, which may affect further size reduction. However, there could be other limiting factors which restrict the production of particles of extremely smaller size.

Redispersibility of the sulfur has been confirmed by checking the PSD after dispersing the dried product in water. Coarsening of the particles is observed in a negligible level. A reduction of about 15% in weight of the particles below 105 μm is observed during redispersion. However, the conglomerated particles still remain below 212 μm in size (Figure 10.4).

By trial and error and the results obtained, a batch size of 400 g of sulfur prills and 800 ml of water with 1 g of SDBS with a run time of 12 minutes is defined for the production of micronized elemental sulfur with ~95% of particles below 37 μm size in the colloidal mill used here.

10.3.2 Particle Size Analysis

Standard test sieves of various sizes are used for size separation of the produced sulfur powder following the procedure outlined in Indian Standard IS: 6655. Weight fractions of size range 0–25 μm, 25–37 μm, 37–63 μm,

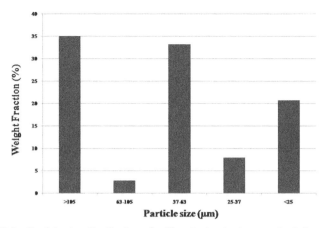

Figure 10.5 Particle size distribution of sulfur obtained using standard sieve analysis.

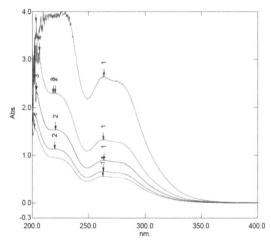

Figure 10.6 UV-absorption spectra of sulfur particles dispersed in methanol at varying concentrations. Absorbance increases with concentration.

63–105 µm, and 105 µm and above were separated. The particle size distribution is shown in Figure 10.5.

10.3.3 UV-Visible Spectroscopy

Sulfur is known to have a UV absorbance at 264 nm when dispersed in ethanol [12]. Various concentrations of size fraction of sulfur particles

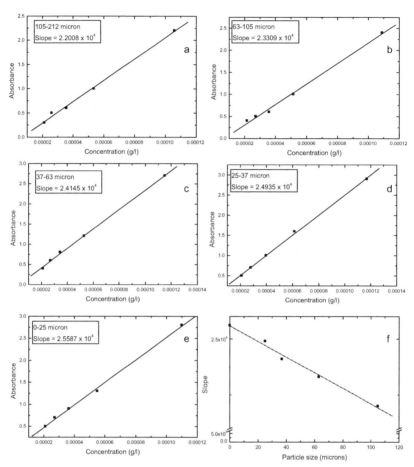

Figure 10.7 (a, b, c, d, e) Concentration versus UV absorbance graphs of sulfur of various particle sizes dispersed in methanol. Slopes of curves increase with decreasing particle size. This relationship is shown in (f), with the straight line relationship of particle size $= (2.6 \times 10^4 - \text{slope})/35$.

separated using the sieves are dispersed in methanol and subjected to ultra-sonication for 5 minutes, and UV-absorption spectra are recorded to obtain the linear increase in absorbance with increasing concentration (Figure 10.6).

Absorbances at wavelength 224 nm for various concentrations of sulfur for each particle size ranges, namely 0–25 μm, 25–37 μm, 37–63 μm, 63–105 μm, and 105 μm and above are plotted and are shown in Figure 10.7a–e. Figure shows that as the particle size decreases, the slope of the absorbance–concentration curve increases. This relationship is depicted

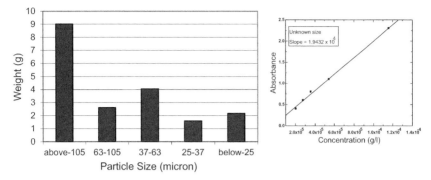

Figure 10.8 Sieve analysis result and UV-absorbance curve of micronized sulfur particles with weight-averaged particle size of 180 μm, of which the UV spectral analysis gave the average particle size as 187 μm.

in Figure 10.7f. A linear curve fit gives the equation:

$$Average\ particle\ size = (2.6 \times 10^4 - Slope)/35 \qquad (10.1)$$

This linear relationship is then applied to an unknown sample to estimate the average particle size (Figure 10.8), which matched with the weight-averaged sieve analysis data. This observation provides a direction towards the application of UV absorption spectrometry for estimating the average particle size of sulfur. However, the time and power parameters of ultrasonication should remain constant.

10.3.4 Fertilizer Application

It is known that sulfur fertilization enhances increase in yield as well as the oil content of groundnut [13]. For testing the applicability as fertilizer, the prepared micronized elemental sulfur is pelletized in the tablet form with suitable calcium lignosulfonate as binding material and bentonite clay as filler material. The tablets of 4 mm diameter and 2.5 mm thickness contained 92% elemental sulfur and <0.5% moisture. A field trial conducted by applying this product to groundnut showed considerable increase in oil content in comparison with control samples.

10.4 Conclusion

Elemental sulfur in micronized form with particle size <100 μm is prepared by wet milling technique in a colloidal mill. Use of dispersing agent has an influence in the final particle size even after redispersing as a fertilizer product. Particles of desirable size range can be achieved by adjusting milling parameters like time, speed, volume, and quantity of dispersant used. Ionic and non-ionic interaction characteristics of the surfactant in the dispersing medium are critical for the selection of surfactant and its concentration.

It can be considered that out of the three surfactants (SDBS, CMC, and SLS), SDBS gives better results when used at a concentration of 0.25% (by weight) of sulfur being micronized. For the mill with similar configuration described, an optimal milling time of 10 minute is suggested for getting particles with D90 = 37 μm.

From the UV absorption results of sulfur particles dispersed in methanol, it can also be proposed that this technique can be applied to estimate the average particle size. Application of micronized sulfur as a fertilizer to groundnut seems to increase oil content in the product.

Acknowledgement

The authors acknowledge the support provided by GSFC Limited for carrying out this research work. Help and support provided by Mr. Ajay Rajput is acknowledged.

References

[1] Massilimov, I. A., Gazyzinovich, M. A., Rifhatovna, S. A., Naile-vich, K. A. and Maratovna, Z. R. Obtaining sulphur nanoparticles from sodium polysulfide aqueous solution, Journal of Chemistry and Chemical Engineering, 2012, 6, p. 233.

[2] Blair, G. J. Sulphur enhanced fertilizer (SEF): A new generation of fertilizers. The proceedings of the international plant nutrition collo-quium XVI. 2009, University of California, Davis. Available at: http://escholarship.org/uc/item/16h5b2dm (verified 7 Feb 2011).

[3] Chapman, S. J. Oxidation of micronized elemental sulphur in soil. Plant and Soil, 1989, 116, p. 69.

[4] Hu, Z. Y., Beaton, J. D., Cao, Z. H. and Henderson, A. Sulfate formation and extraction from red soil treated with micronized elemental sulphur fertilizer and incubated in closed and open systems. Communications in Soil Science and Plant Analysis, 2002, 33, p. 11.

[5] Kemper, W. D. and Rosenau, R. C. Aggregate stability and size distribution. In: Methods of soil analysis. Part 1. Physical and Mineralogical Methods-Agronomy Monograph No. 9 (2nd Edition), 1986, p. 425.

[6] Komar, P. D. and Cui, B. The analysis of grain size measurements by sieving and settling tube techniques. Journal of Sedimentary Petrology, 1984, 54, p. 603.

[7] Shiffrin, K. S. and Tonna, G. Inverse problems related to light scattering in the atmosphere and ocean. Advances in Geophysics, 1993, 34, p. 175.

[8] Bertero, M., De Mol, C. and Pike, E. R. Particle size distributions from spectral turbidity: A singular-system analysis. Inverse Problems, 1986, 2, p. 247.

[9] Wang, J. and Ross Hallet, F. Spherical particle size determination by analytical inversion of the UV-visible–NIR extinction spectrum. Applied Optics, 1996, 35, p. 193.

[10] Tscharnuter, W. Photon correlation spectroscopy in particle sizing. In: Meyers, R. A. (Ed.) Encyclopedia of analytical chemistry, 2000, p. 5469.

[11] Turganbay, S., Aidrova, S. B., Bekturganova, N. E., Sheng Li, C., Musabekov, K. B., Kumargalieva, S. and Toshtay, K. Nanoparticles of sulphur as fungicidal products for agriculture. Eurasian ChemTech Journal, 2012, 14, p. 313.

[12] Heatley, N. G. and Page, E. J. Estimation of elemental sulphur by ultraviolet absorption. Analytical Chemistry, 1952, 24(11), p. 1854.

[13] Haneklaus, S., Bloem, E. and Schnug, E. The global sulphur cycle and its links to plant environment. In: Abrol, Y. P. and Ahmad, A. (eds) Sulphur in plants. Springer, Dordrecht, 2003.

11

Measurements of Temporal Fluctuations of Magnetization in Alkali Vapor and Applications

Maheswar Swar[*]**, Dibyendu Roy, Dhamodaran Dhanalakshmi, Saptarishi Chaudhuri, Sanjukta Roy and Hema Ramachandran**

Raman Research Institute, Sadashivanagar, Bangalore, 560080, India
E-mail: mswar@rri.res.in, droy@rri.res.in, dhana@rri.res.in, srishic@rri.res.in, sanjukta@rri.res.in, hema@rri.res.in
[*]Corresponding Author

We describe a relatively non-invasive optical spectroscopy technique, known as spin noise spectroscopy (SNS), to probe spontaneous magnetization fluctuations in an alkali atomic vapor at thermal equilibrium. The atomic magnetization at a non-zero temperature fluctuates about its equilibrium value, and such magnetization (spin) noise imparts fluctuation in the measured Faraday rotation of a far-detuned, linearly polarized probe laser beam passing through the vapor. This Faraday rotation fluctuation reveals the intrinsic spin dynamics of the electrons in the alkali vapor. We present a systematic experimental study of the SNS realized in thermal rubidium vapor and illustrate how one can measure various magnetic and chemical properties of the sample with minimal perturbation. We also demonstrate the applicability of this spectroscopy technique in measuring the spin populations and their dynamics in different hyperfine and magnetic levels by optically pumping the atomic system.

11.1 Introduction

The fluctuation-dissipation theorem [1] in statistical physics relates the linear response function of a system measured by applying a small perturbation to the system's spontaneous fluctuation properties at thermodynamic equilibrium. Most spectroscopy techniques, such as nuclear magnetic resonance (NMR) probe the linear response function by preparing a system in a non-equilibrium state. Spin noise spectroscopy (SNS) [2, 3] has been developed as an alternative method to detect the statistical fluctuation in magnetization of a system at thermal equilibrium. The SNS can reveal the dynamical spin properties, such as spin relaxation times, of the system with a minimal perturbation.

For a system with an ensemble of electron spins at thermal equilibrium, the magnetization of the system fluctuates about its equilibrium value. This causes spontaneous relaxation of spins from one magnetic sub-level to another. The time-averaged value of the magnetization, $\langle M(t) \rangle_{T \to \infty}$ (T is the total averaging time), along any arbitrary quantization axis is zero for a paramagnet in the absence of an external magnetic field. However, the variance of magnetization is still non-zero [4]. An off-resonant and linearly polarized laser beam passing through such a paramagnetic sample can detect this magnetization (spin) noise as a fluctuation in the time-resolved Faraday rotation ($\theta_F(t)$) of its polarization vector. In the presence of a constant magnetic field ($B_\perp$) being perpendicular to the light propagation, the magnetization fluctuation along the light propagation direction (z-axis) can be found from the Bloch equations as $\langle M_z(t)M_z(0) \rangle \propto \cos(\omega_L/2\pi)e^{-\frac{t}{T_2}}$, where $\omega_L/2\pi \, (= |g|\mu_B B_\perp/h)$ is the Larmor frequency and T_2 is the transverse spin relaxation time along the z-axis. Here, g is the Lande g-factor of the ground states, μ_B is the Bohr Magneton, and $\hbar$ is the reduced Planck's constant.

The experimentally probed Faraday rotation fluctuation $\langle \theta_F(t)\theta_F(0) \rangle$ is a measure of $\langle M_z(t)M_z(0) \rangle$. The Fourier transform of $\langle \theta_F(t)\theta_F(0) \rangle$ in frequency domain is the desired spin noise (SN) spectrum. The line-shape of this SN spectrum is a Lorentzian centered at $\omega_L/2\pi$ whose full width at half maxima is given by $1/T_2$. The amplitude of the SN can be written as [5]:

$$\sqrt{\langle (\Delta_F)^2 \rangle} \propto \frac{I_p}{|\delta|}\sqrt{\frac{n_0 l}{A}} \qquad (11.1)$$

where n_0 is the density of atoms; l is the length of the atomic vapor in a cell; A, I_p, and δ are the cross-sectional area, the intensity, and the detuning of the probe beam, respectively.

The SNS technique has been implemented in experiments with hot alkali atomic vapors [6–9], semiconductor heterostructures, and quantum dots [10–13]. Here, we describe our effort to perform and investigate unexplored aspects of SNS in thermal rubidium (Rb) vapor.

11.2 Methodology and Experimental Set-up

A linearly polarized, off-resonant probe laser beam with a tunable frequency is sent through a 20 mm long vapor cell containing enriched ^{87}Rb vapor. The probe beam is focused inside the atomic medium to a Gaussian $1/e^2$ waist size of 45 μm. The frequency of the probe was measured using a commercial wavelength meter with a relative accuracy of ± 1 MHz. The vapor cell is connected to a controllable heater to study the spin noise with different atom number density. The vapor cell is filled with neon buffer gas with a partial pressure of 200 millibar. It makes the atomic medium diffusive which increases the transit time of thermal rubidium atoms through the probe beam allowing for measurement of the intrinsic atomic spin fluctuation with a good signal to noise ratio.

A constant magnetic field ($B_\perp$) being perpendicular to the propagation direction of the probe beam is applied on the atoms. The polarization of the transmitted probe beam is measured using a polarization sensitive detection setup comprising of a half-wave plate (HWP) and a polarizing cube beam splitter (PBS) as shown in Figure 11.1. The s- and p-polarized components from the two arms of the PBS are fed to two detector ports of a balanced photodetector with 80 MHz bandwidth and 25 dB common mode rejection ratio. The output of the balanced detector is directly connected to a spectrum analyzer.

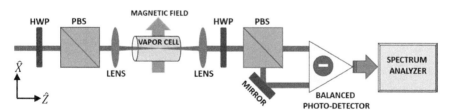

Figure 11.1   Sketch of SNS setup to detect spin noise. The probe beam from the left passes through a half-wave plate (HWP), a polarizing beam splitter (PBS), and a focusing lens before entering the vapor cell. The polarization of the transmitted beam is detected by a polarimetric setup, a balanced photodetector, and a spectrum analyzer.

11.3 Results

In Figure 11.2, two typical spin noise power spectra at a low $B_\perp$ (= 5.63 G) are presented. The spectrum analyzer is set on continuous averaging mode for 2 minutes to record both these spectrums. The photon shot noise background is subtracted from these noise spectrums. The noise signal shown using red dots corresponds to the signal recorded with a probe beam having $\delta = -10.4$ GHz and a power of 200 µW. In the same graph, a noise signal is shown using blue dots recorded with a probe beam having $\delta = -15.1$ GHz and a power of 100 µW. The variation of the noise signal strengths is attributed to the dependence of the SN amplitude on the probe beam intensity and detuning as presented in Equation (11.1).

The noise signal shown with red dots has two distinct noise peaks. The stronger (weaker) signal peaked at 3.94 MHz (2.63 MHz) is identified as the spin noise signal due to spin fluctuation among the intra-hyperfine magnetic sublevels, $\Delta F = 0$, $\Delta m_F = \pm 1$ (see Figure 11.3) of ^{87}Rb (^{85}Rb). The weaker signal signifies the trace amount of ^{85}Rb atoms in the vapor cell enriched with ^{87}Rb and manifests the merit of SNS for precision measurement of isotope abundance.

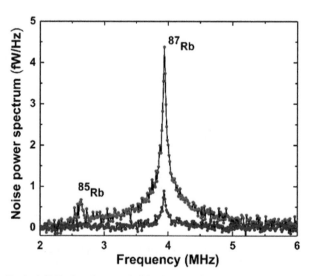

Figure 11.2   Typical SNS signals recorded in the experiment. Averaged SN power spectrum with $B_\perp = 5.63$ G at a cell temperature of 100°C, $\delta = -10.4$ GHz, and the probe beam power = 200 µW (red dot with black line) and $\delta = -15.1$ GHz and the probe beam power = 100 µW (blue dot with black line).

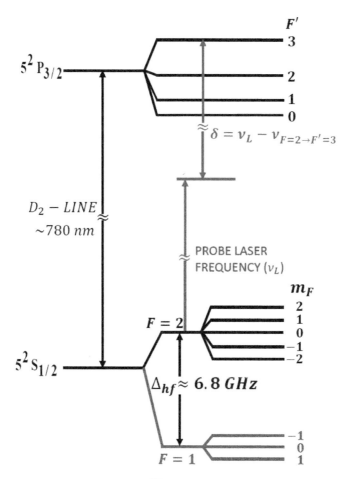

Figure 11.3 Energy level diagram for [87]Rb atoms. The spin fluctuation between different Zeeman sub-levels of the hyperfine F = 1, 2 states causes the polarization fluctuation of a probe beam whose frequency is detuned by δ from F = 2 → F' = 3 transition. The laser frequency is ν_L and the atomic transition frequency (F = 2 → F' = 3) is $\nu_{F=2 \to F'=3}$. We define the detuning δ as ($\nu_L - \nu_{F=2 \to F'=3}$). The Zeeman sub-levels of F = 1, 2 states in the presence of a magnetic field are shown above.

In Figure 11.4, four typical SN spectra are shown by varying $B_\perp$ between 3 and 9 G. In each of these spectra, two separate noise peaks corresponding to [87]Rb and [85]Rb are observed. By measuring the SN peak positions, the Lande g-factor of the ground hyperfine states ($|g_F|$) is precisely estimated to be 0.500(1) for [87]Rb and 0.333(1) for [85]Rb which agree nicely with

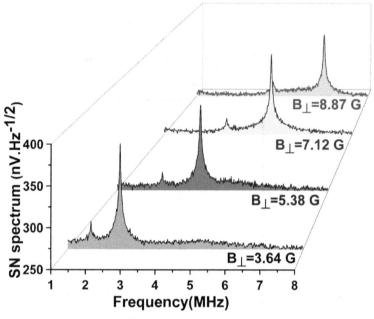

Figure 11.4   Averaged SN spectra with $\delta = -10.6$ GHz for various $B_\perp$ ranging between 3 and 9 G. Probe power = 150 μW, vapor cell temperature = 105°C.

the theoretical values. The ratio of the total integrated SN spectrum around individual peaks in Figure 11.4 gives the abundance ratio of ^{87}Rb: ^{85}Rb $\approx$ 7:1 in the vapor cell. These data show that the SNS can also be used as a sensitive technique to detect isotope abundance with high precision.

11.3.1 Precision Magnetometry Using Spin Noise Spectroscopy

In this section, we demonstrate the applicability of spin noise spectroscopy in precision measurement of magnetic field.

In the presence of a high $B_\perp$ (typically higher than 20 G), the individual spin noise peak splits into multiple peaks due to non-linear Zeeman effect as shown in Figure 11.5.

The energy of a particular Zeeman sub-level is a function of $B_\perp$ which is given by the generalized Breit-Rabi formula for alkali atoms [14]:

$$E_{F,m_F} = -\frac{h\Delta_{\text{hf}}}{2(2I+1)} + g_I\mu_B B_\perp m_F \pm \frac{h\Delta_{\text{hf}}}{2}\sqrt{1 + \frac{4m_F}{2I+1}x + x^2}$$

$$(11.2)$$

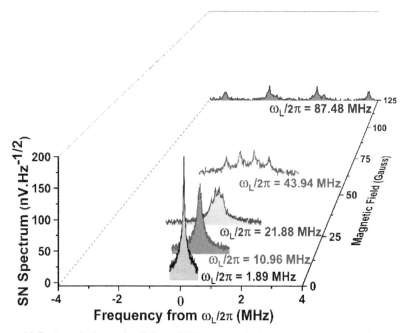

Figure 11.5 Broadening and splitting of SN spectrum in the presence of progressively higher $B_\perp$ according to the Breit-Rabi formula. Here, we show SNS signals corresponding to $B_\perp =$ 2.70 G (black), 15.66 G (red), 31.26 G (blue), 62.77 G (magenta), and 124.97 G (brown). The SN spectra are plotted relative to the central Larmor precession frequency. Probe power $= 400$ μW, $\delta = -10.6$ GHz, and vapor cell temperature $= 105°$C.

where Δ_{hf} is the zero-field hyperfine separation between the ground states, g_I is the nuclear g-factor, $x = \frac{(g_J - g_I)\mu_B B_\perp}{h\Delta_{hf}}$ and $\pm$ refer to the $F = I \pm 1/2$ ($I = 3/2$ for ^{87}Rb) hyperfine states.

Using Equation (11.2), and typical spin noise spectrum at high $B_\perp$ as shown in the insets on Figure 11.6, we estimate the value of $B_\perp$ for each spectrum. We have used the CODATA value of g_I [15] and Δ_{hf} [16] for this estimation. This way of estimating the external magnetic field has the advantage that we can completely avoid the systematic errors arising due to the effects of stray magnetic field in the laboratory, calibration errors in the current supply. We have recorded a large number of these spectra at various $B_\perp$. In Figure 11.6, we finally plot the frequency separation between the measured noise peaks as a function of the external magnetic field. These data are fitted with Equation (11.2) keeping the value of Δ_{hf} as a free parameter. From the fits we estimate the error in the zero-field hyperfine separation, Δ_{hf}

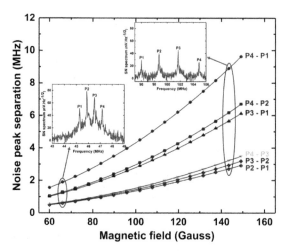

Figure 11.6 Precision measurement of magnetic field using spin noise spectroscopy. Separation between spin noise peaks is plotted against the external magnetic field. The data are fitted with generalized Breit-Rabi formula keeping Δ_{hf} as a free parameter. In this plot, P1...P4 refers to the position of the spin noise peaks (corresponding to F = 2 Zeeman sub-levels) in frequency as can be understood from the raw data shown in the insets.

to be less than 0.2%. In other words, for any subsequent measurements we can estimate the magnetic field value with a similar 0.2% error.

At reasonably higher $B_\perp$ (>150 G), the difference between the transitions $(2,1) \leftrightarrow (2,0)$ and $(1,1) \leftrightarrow (1,0)$ (and also between $(2,0) \leftrightarrow (2,-1)$ and $(1,0) \leftrightarrow (1,-1)$) is more than the width of the individual SN peaks due to the nuclear spin contribution which is the second term in Equation (11.2). In this case, we can resolve the SN spectrum between all available Zeeman coherences. In Figure 11.7, six distinct SN peaks are shown at $B_\perp = 156.02$ G. Measuring the spacing between $(2,1) \leftrightarrow (2,0)$ and $(1,1) \leftrightarrow (1,0)$, the value of the nuclear g-factor (g_I) is precisely obtained as $-0.00100627(2558)$ which is in excellent agreement with other precision measurements of the same quantity.

The SNS can act as a relatively non-invasive technique if the photon scattering by the atoms is negligibly small. Both photon scattering rate (I_p/δ^2) and SN amplitude $(I_p/|\delta|)$ depend on I_p and δ of the probe beam. Therefore, the polarization fluctuation of a probe beam can reveal the intrinsic spin noise of a sample at thermal equilibrium only for a large detuning and a relatively low power of the probe beam. In this non-invasive regime, we extract the transverse spin relaxation time $T_2 \approx 25$ μS.

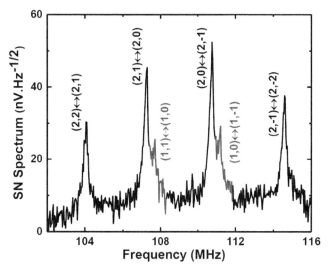

Figure 11.7 Resolved SN spectrum between all intra-hyperfine Zeeman sub-levels in [87]Rb. The measurement is taken at $B_\perp = 156.02$ G.

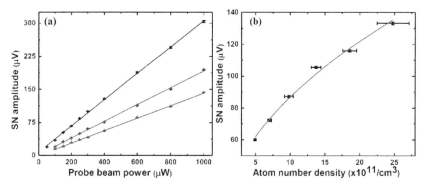

Figure 11.8 (a) Integrated SN amplitude as a function of total probe beam power for three different detuning of the probe beam, $\delta = +9$ GHz (black), $\delta = +15.1$ GHz (red), and $\delta = +20.1$ GHz (blue). The solid lines are the linear fit to the data. (b) Dependence of SN amplitude on atom number density. The line is a power law fit of the data with exponent 0.49 (± 0.02).

In Figure 11.8a, we show the linear dependence of the measured SN amplitude on the probe beam intensity as discussed earlier (Equation 11.1). In Figure 11.8b, the SN amplitude is plotted against the atom number density in the vapor cell. We observe the expected scaling of $\sqrt{n_0}$ as predicted in Equation (11.1).

11.3.2 Measurement of Spin Noise in a Driven System

We further explored the spin fluctuations of the atomic system in a pump-probe arrangement in order to measure the changes in spin dynamics and population difference in different hyperfine ground states in an optically pumped atomic system. Spin noise measurement in such a non-equilibrium system reveals the information about its non-linear properties. A similar measurement of spin noise in atomic system where the application of weak radio frequency (RF) field leads to beyond linear response of the system is reported in Ref. [17]. In our experiment, we introduce an on-resonance pump beam with high intensity, nearly co-propagating with the probe beam, to optically pump all the atoms to any one of the ground hyperfine levels F = 1 or F = 2. We take care that no part of this pump beam goes through our polarimetric detection setup. The spin noise probe beam explores the spin dynamics information of the atom optically pumped to the individual hyperfine levels. In Figure 11.9b, left side shows the spin dynamics between all the possible intra-hyperfine magnetic sub-levels in F = 1 and F = 2 ground state manifold. Hence, we observe all six possible spin noise peaks in this spectrum.

When we use a pump beam with 50 times saturation intensity, tuned from F = 1 to F' = 2 transitions (see top right-side energy diagram in Figure 11.9), almost all the atoms in the detection zone are optically pumped to F = 2 ground hyperfine state. The orange colored arrows in the same graph show the possible dipole allowed optical transitions and spontaneous emission pathways of the ^{87}Rb atoms initially distributed in ground hyperfine states F = 1 and F = 2. Therefore, the SNS probe detects the spin coherence between the magnetic sub-levels belonging to F = 2 hyperfine level. As expected, we observe four (instead of six) spin noise peaks in this case as shown in Figure 11.9a. Similarly, when we optically pump most of the atoms to F = 1 ground hyperfine state (see bottom right-side energy diagram in Figure 11.9), only two spin noise peaks are observed as shown in the Figure 11.9c. This observation provides information about the hyperfine state populations without significantly altering the initial populations. In a large number of physical systems (such as atomic gravimeters [18] and atomic clocks [19]) where information on real-time hyperfine state population is required, this technique of spin noise spectroscopy can be extremely useful.

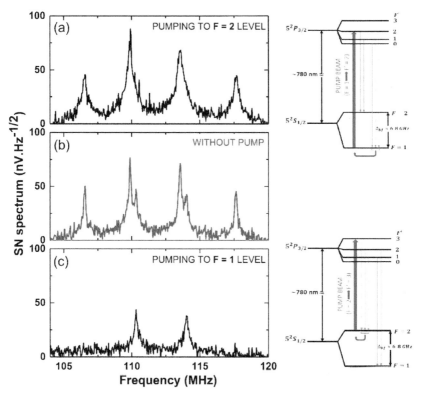

Figure 11.9 Measurement of spin population dynamics in optically pumped atomic system. (a) A typical spin noise spectrum from an ensemble of ^{87}Rb atoms optically pumped to F = 2 ground hyperfine state. (b) A corresponding spectrum with thermal equilibrium population distribution between F = 1 and F = 2 states (no optical pumping). (c) Spin noise spectrum is recorded from an optically pumped vapor where all the atoms are in F = 1 ground hyperfine state.

11.4 Conclusion

We have reported here a detailed experimental study of the optical spectroscopy of electron spin noise in a thermal rubidium atomic vapor. The applicability of this technique in precision measurements of isotope abundance, hyperfine constant, nuclear magnetic moment, and spin relaxation time is demonstrated. Moreover, we also presented how the spin noise spectroscopy can be used to infer the real-time spin population dynamics in different hyperfine states in neutral atoms.

Acknowledgement

The authors acknowledge the contribution of Meena M.S. for the help with electronics and the mechanical workshop of Raman Research Institute for the hardware development.

Author's Contribution

MS, DD, and SC designed the experiments and recorded the data. MS, DR, SR, and SC analyzed the data and prepared the manuscript. All authors took part in science discussions related to finalizing the manuscript.

References

[1] Kubo, R., The fluctuation-dissipation theorem, Rep. Prog. Phys., **29**, 255284 (1966).

[2] Zapasskii, V. S., Spin-noise spectroscopy: From proof of principle to applications, Adv. Opt. Photon., **5**, 131–168 (2013).

[3] Sinitsyn, N. A. and Pershin, Y. V., The theory of spin noise spectroscopy: A review, Rep. Prog. Phys., **79**, 106501 (2016).

[4] Bloch, F., Nuclear induction, Phys. Rev., **70**, 460–474 (1946).

[5] Mihaila, B., Crooker, S. A., Rickel, D. G., Blagoev, K. B., Littlewood, P. B. and Smith, D. L., Quantitative study of spin noise spectroscopy in a classical gas of ^{41}K atoms, Phys. Rev. A, **74**, 043819 (2006).

[6] Aleksandrov, E. B. and Zapasskii, V. S., Magnetic resonance in the Faraday-rotation noise spectrum, Zh. Eksp. Teor. Fiz., **81**, 132–138 (1981).

[7] Sleator, T., Hahn, E. L., Hilbert, C. and Clarke, J., Nuclear-spin noise, Phys. Rev. Lett., **55**, 1742 (1985).

[8] Crooker, S. A., Rickel, D. G., Balatsky, A. V. and Smith, D. L., Spectroscopy of spontaneous spin noise as a probe of spin dynamics and magnetic resonance, Nature, **431**, 49 (2004).

[9] Roy, D., Yang, L, Crooker, S. A. and Sinitsyn, N. A., Cross-correlation spin noise spectroscopy of heterogeneous interacting spin systems, Sci. Rep., **5**, 9573 (2015).

[10] Oestreich, M., Römer, M., Haug, R. J. and Hägele, D., Spin noise spectroscopy in GaAs, Phys. Rev. Lett., **95**, 216603 (2005).

[11] Römer, M., Hübner, J. and Oestreich, M., Spin noise spectroscopy in semiconductors, Rev. Sci. Instrum., **78**, 103903 (2007).

[12] Römer, M., Hübner, J. and Oestreich, M., Spatially resolved doping concentration measurement in semiconductors via spin noise spectroscopy, Appl. Phys. Lett., **94**, 112105 (2009).

[13] Crooker, S. A., Cheng, L. and Smith, D. L., Spin noise of conduction electrons in n-type bulk GaAs, Phys. Rev. B., **79**, 035208 (2009).

[14] Steck, D.A., Rubidium 87 D Line Data, 2.1.5, 13 January 2015.

[15] Arimondo, E., Inguscio, M. and Violino, P., Experimental determinations of the hyperfine structure in the alkali atoms, Rev. Mod. Phys., **49**, 31 (1977).

[16] Bize, S., Sortais, Y., Santos, M. S., Mandache, C., Clairon, A. and Salomon, C., High-accuracy measurement of the ^{87}Rb ground-state hyperfine splitting in an atomic fountain, Eur. Phys. Lett., **45**, 558 (1999).

[17] Glasenapp, P., Sinitsyn, N. A., Yang, L., Rickel, D. G., Roy, D., Greilich, A., Bayer, M. and Crooker, S. A., Spin noise spectroscopy beyond thermal equilibrium and linear response, Phys. Rev. Lett., **113**, 156601 (2014).

[18] Bertoldi, A., Lamporesi, G., Cacciapuotia, L., Angelisb, M. de, Fattori, M., Tino, G. M. et al., Atom interferometry gravity-gradiometer for the determination of the Newtonian gravitational constant G, Eur. Phys. J. D, **40**, 271 (2006).

[19] Wynands, R. and Weyers, S., Atomic fountain clocks, Metrologia, **42**, S64 (2005).

12

Lie Algebraic Approach to Molecular Spectroscopy: Diatomic to Polyatomic Molecules

V. K. B. Kota

Physical Research Laboratory, Ahmedabad 380009, India
E-mail: vkbkota@prl.res.in

Interacting dipole (p) bosons along with scalar (s) bosons, based on the ideas drawn from the interacting boson model of atomic nuclei, led to the development of the vibron model based on $U(4)$ spectrum generating algebra for diatomic molecules. The $U(4) \supset SO(4) \supset SO(3)$ algebra generates rotation–vibration spectra. Extending this to two coupled $SO(4)$ algebras and three coupled $SO(4)$ algebras describes triatomic- and four-atomic molecules respectively. Similarly, appropriately coupled $U(2) \supset SO(2)$ algebras will describe the stretching vibrations, with proper point group symmetries, in polyatomic molecules. In addition, coupled $U(3)$ algebras describe coupled benders. The Lie algebraic approach to molecular spectroscopy is briefly reviewed along with a list giving future directions and some results for order-chaos transitions and partition functions.

12.1 Introduction

Quantizing the relative co-ordinate in diatomic molecules gives rise to a description of vibrational–rotational spectra in terms of interacting dipole (π) bosons with $\ell = 1^-$. The π bosons along with scalar (s) bosons, based on the ideas drawn from the interacting boson model of atomic nuclei [1, 2], led to the development of the vibron model for diatomic molecules with $U(4)$ spectrum generating algebra (SGA) [3, 4]. The $SO(4)$ subalgebra in

$U(4) \supset SO(4) \supset SO(3)$ generates rotation–vibration spectra; $SO(3)$ generates angular momentum. Extension with two coupled $U(4) \supset SO(4)$ algebras describes stretching and bending vibrations in triatomic molecules. Similarly, three coupled $U(4) \supset SO(4)$ algebras describe four-atomic molecules. Continuing this to several coupled $SO(4)$ algebras will in principle describe polyatomic molecules but these algebras will become unwieldy for molecules with 5 or more atoms. Then, an alternative is to use coupled $U(2) \supset SO(2)$ algebras. This, along with a Majorana force, will allow for incorporating the point group symmetries of polyatomic molecules within the Lie algebraic approach and describe, for example, stretching vibrations in a variety of polyatomic molecules [5, 6]. In addition, it is also possible to use the more complicated, but simpler than $U(4) \supset SO(4)$, coupled $U(3) \supset SO(3)$ algebras [7–9]. Going beyond these and using the ideas from the interacting boson-fermion model of atomic nuclei [10–12], Lie algebraic approach is also shown to describe molecular electronic spectra [13]. In this article, we will give an overview of these developments in the Lie algebraic approach to molecular spectroscopy with emphasis on group theoretical aspects. Now we will give a preview.

Section 12.2 gives the results of $SO(4)$ algebra for diatomic molecules. Similarly, Section 12.3 gives the results of coupled $SO(4)$ algebras for triatomic and four-atomic molecules. Section 12.4 is on coupled $SU(2) \supset SO(2)$ algebras for polyatomic molecules. Section 12.5 gives in some detail $U(3)$ algebra for bending vibrations and coupled benders. Section 12.6 gives conclusions along with a list giving future directions. Finally, discussed briefly in Section 12.5 and in an Appendix are applications to order-chaos transitions, quantum phase transitions (QPT) and partition functions.

12.2 *SO*(4) Algebra for Diatomic Molecules

Quantizing the relative co-ordinate $\vec{r}$ between the two atoms of a diatomic molecule, we have the vector boson (π boson) with $\ell = 1^{-}$; $\pi_{\mu}^{\dagger} = (r_{\mu} - ip_{\mu})/\sqrt{2}$ and $\pi_{\mu} = (r_{\mu} + ip_{\mu})/\sqrt{2}$. Now, introducing s bosons ($\ell = 0^{+}$) and demanding that the total number (N) of π and s bosons is conserved, we have the vibron model with $U(4)$ spectrum generating algebra (SGA). The $U(4)$ is generated by the 16 one-body operators $\pi_{\mu}^{\dagger}\pi_{\mu'}$, $s^{\dagger}s$, $\pi_{\mu}^{\dagger}s$, $s^{\dagger}\pi_{\mu'}$. In angular momentum coupled representation, introducing $\tilde{\pi}_{\mu} = (-1)^{1+\mu}\pi_{-\mu}$ the number operator for π bosons is $n_{\pi} = \sqrt{3}(\pi^{\dagger}\tilde{\pi})^{0}$ and, similarly, $n_s = s^\dagger s$. They will give the number of π bosons N_π and s bosons N_s with $N = N_\pi + N_s$. The angular momentum operator $L_\mu^1 = \sqrt{2}(\pi^\dagger \tilde{\pi})_\mu^1$. Using the commutation relations between the $U(4)$ generators, it is easy to see that $U(4) \supset SO(4) \supset SO(3) \supset SO(2)$ where $SO(4)$ is generated by the 6 operators L_μ^1 and $D_\mu^1 = i(\pi^\dagger s + s^\dagger \tilde{\pi})_\mu^1$, $SO(3)$ by L_μ^1 and $SO(2)$ by L_0^1. Let us add that it is also possible to have another $SO(4)$ algebra (called $\overline{SO(4)}$) generated by L_μ^1 and $\mathcal{D}_\mu^1 = (\pi^\dagger s - s^\dagger \tilde{\pi})_\mu^1$. We will not consider $\overline{SO(4)}$ any further in this article except in the Appendix. The quantum numbers [called irreducible representations (irreps) in the representation theory of Lie algebras] of $U(4)$, $SO(4)$ and $SO(3)$ are N, ω and L, respectively. The M quantum number of $SO(2)$ is trivial, and it is dropped from now on as we deal with only L scalar Hamiltonians. The $N \to \omega \to L$ irrep reductions are easy to identify using pairing algebra in nuclear physics and also using many other approaches [3, 4, 14]. Then we have, $N \to \omega = N, N - 2, N - 4, \ldots, 0$ or 1 and $\omega \to L = 0, 1, 2, \ldots, \omega$. Using only the quadratic Casimir invariants, the $U(4)$ Hamiltonian [assuming one plus two-body in nature and preserving N and L] for diatomic molecules (H_{d-m}) is:

$$\begin{aligned} H_{d-m} &= E_0 + \alpha C_2(SO(4)) + \beta C_2(SO(3)) \\ &= E_0 + \alpha\left(L^2 + D^2\right) + \beta L^2. \end{aligned} \tag{12.1}$$

Here, E_0 is a function of N. Using the known formulas for the Casimir invariants will give $E = E_0 + \alpha\omega(\omega + 2) + \beta L(L + 1)$; note that $\langle C_2(SO(4))\rangle^{N,\omega,L} = \omega(\omega+2)$ [2]. Changing ω into the vibrational quantum number $v = (N - \omega)/2$ will give the energy formula:

$$\begin{aligned} &E = E_0' - 4\alpha(N + 2)(v + \tfrac{1}{2}) + 4\alpha(v + \tfrac{1}{2})^2 + \beta L(L + 1); \\ &v = (N - \omega)/2 = 0, 1, 2, \ldots, \left[\tfrac{N}{2}\right] \text{ or } \left[\tfrac{N-1}{2}\right], \\ &v \to L = 0, 1, 2, \ldots, (N - 2v). \end{aligned} \tag{12.2}$$

Therefore, with N large, $\alpha < 0$ and $\beta > 0$, the $SO(4)$ algebra generates rotation–vibration spectrum as seen clearly, for example, in H_2 molecule in its electronic ground state (here $N \sim 31$ and this follows from the observed v_{max} value). In fact $SO(4)$ represents rigid molecules (this can be derived from the Morse oscillator) and the other limit $U(4) \supset [SU(3) \supset SO(3)] \oplus U(1)$ is for non-rigid molecules [3], see Appendix. It is important to recognize that Equation (12.2) is similar to the well-known Dunham expansion [15].

12.3 Coupled *SO*(4) Algebras for Triatomic and Four-atomic Molecules

Let us start with triatomic molecules. Now there are two relative co-ordinates and associating $U(4)$ SGA to each of these, and the SGA for triatomic molecules is $U_1(4) \oplus U_2(4)$. This SGA admits large number of subalgebras but the most important are: (i) local basis generated by $U_1(4) \oplus U_2(4) \supset SO_1(4) \oplus SO_2(4) \supset SO_{12}(4) \supset SO(3)$; (ii) normal basis generated by $U_1(4) \oplus U_2(4) \supset U_{12}(4) \supset SO_{12}(4) \supset SO(3)$. In the local basis, the two $U_{i=1,2}(4)$ algebras give boson numbers N_1 and N_2 and similarly the two $SO_{i=1,2}(4)$ give ω_1 [or $v_1 = (N_1 - \omega_1)/2$] from N_1 and ω_2 [or $v_3 = (N_2 - \omega_2)/2$] from N_2, respectively. The $SO_{12}(4)$ irreps are (τ_1, τ_2) and they are generated by the so-called Kronecker product of ω_1 and ω_2. This then gives (see for example [3, 14] for the Kronecker products):

$$(\tau_1, \tau_2) = \sum_{\alpha, \beta} (\omega_1 + \omega_2 - \alpha - \beta, \alpha - \beta);$$

$$\alpha = 0, 1, \ldots, \min(\omega_1, \omega_2), \quad \beta = 0, 1, \ldots, \alpha. \tag{12.3}$$

Similarly, the reduction of $(\tau_1, \tau_2) \to L$ follows from the recognition that $SO(4)$ is isomorphic to $SO(3) \otimes SO(3)$ and the two $SO(3)$'s are labeled by $J_1 = (\tau_1 + \tau_2)/2$ and $J_2 = (\tau_1 - \tau_2)/2$; $\tau_1 \geq \tau_2$. Then, the simple angular momentum coupling rule gives $J_1 \times J_2 \to L$. The final result is:

$$L = 0^+, 1^-, 2^+, \ldots, \tau_1^\pi; \quad \text{for } \tau_2 = 0 \text{ and } \pi = (-1)^{\tau_1}$$

$$L = \tau_2^\pm, (\tau_2 + 1)^\pm, \ldots, (\tau_1)^\pm; \quad \text{for } \tau_2 \neq 0. \tag{12.4}$$

More conventional notation for (τ_1, τ_2) is to use $v_2^{\ell_2}$ with $v_2 = N_1 + N_2 - 2v_1 - 2v_3 - \tau_1$ and $\ell_2 = \tau_2$. Using Equation (12.3), we have, $v_2 = 0, 1, 2, \ldots, 2*\min(N_1 - 2v_1, N_2 - 2v_3)$ and $\ell_2 = v_2, v_2 - 2, \ldots$ 0 or 1. Note that $\ell_2 = 0, 1, 2, 3, 4, \ldots$ are in spectroscopic notation $\Sigma, \Pi, \Delta, \Phi, \Gamma$ and so on. Adding the L and D operators from the two $SO(4)$'s will give the quadratic Casimir invariant $L_{12}^2 + D_{12}^2$ of $SO_{12}(4)$, and its eigenvalues in (τ_1, τ_2) irreps are $[\tau_1(\tau_1 + 2) + \tau_2^2]$. Now, using $H_{t-m} = E_0 + a_1 C_2(SO_1(4)) + a_2 C_2(SO_2(4)) + a_{12} C_2(SO_{12}(4)) + a_3 L_{12}^2$ will give a formula exactly similar to the Dunham expression:

$$E(v_1 v_2^{\ell_2} v_3 L) = E_0' + \sum_i \alpha_i(v_i + d_i) + \sum_i \beta_i(v_i + d_i)^2$$

$$+ \sum_{i<j} \gamma_{ij}(v_i + d_i)(v_j + d_j) + g_{12}\ell_2^2 + hL(L+1). \tag{12.5}$$

where $d_i = 1/2$ for v_1 and v_3 and 1 for v_2. For linear triatomic molecules Equation (12.5) is good. However, for bent molecules the projection quantum number k (same as ℓ_2 but ℓ_2 is used for linear molecules) can take any value and different k states are expected to be degenerate. Here, we define (v_2', k) via $\tau_1 = N_1 + N_2 - 2v_1 - 2v_3 - 2v_2' - k$ and $\tau_2 = k$. Then, $v_2' = 0, 1, 2, \ldots$ and $k = 0, 1, 2, 3, \ldots$ for any v_2'. To obtain k degeneracy, we need to consider $\overline{C_2(SO_{12}(4))} = \sqrt{|L \cdot D|^2}$, and its eigenvalues in the $(\tau_1 \tau_2)$ irreps are $\tau_2(\tau_1 + 1)$. Therefore, adding $2a_{12}\overline{C_2(SO_{12}(4))}$ to H_{t-m} will give $a_{12}(\tau_1 + \tau_2)(\tau_1 + \tau_2 + 2)$ and then E is independent of the k quantum number.

Turning to the normal mode basis, it is easy to identify that the $U_{12}(4)$ irreps will be $\{N_a, N_b\} = \{N_1 + N_2 - n, n\}$ where $n = 0, 1, 2, \ldots,$ $\min(N_1, N_2)$. The $\{N_a, N_b\} \to (\tau_1, \tau_2)$ reductions can be written down but they involve the more complicated 'multiplicity' label, see for example [3, 14]. One usefulness of $U_{12}(4)$ is that it can be used to mix local basis states and in reality, for describing linear or bent molecules some mixing is essential. The Majorana interaction M_{12}, which is related to $C_2(SU_{12}(4))$ in a simple manner and has a proper physical meaning [2], is added to H for generating mixing. Inclusion of M_{12} term in H_{t-m} is similar to Darling-Dennison coupling between the local modes v_1 and v_3 [3]. In addition, also a Fermi coupling term F_{12} and higher order terms in Casimir operators are added to H. With these, good agreements with data (within 1–5 cm^{-1}) are obtained for many triatomic molecules such as H_2O, SO_2, CO_2, HCN, OCS, H_2S, D_2O, N_2O and in some with different isotopes (ex: $C^{12}O_2$, $C^{13}O_2$, H_2O^{16}, H_2O^{18}). Depending on the molecule, $N_1 = N_2$ or $N_1 \neq N_2$. Also, in all the cases the value of N_i is quite large. Besides comparing spectra, the algebraic approach also allows for calculating intensities of vibrational excitations, see [3, 16, 17] for details. All these extend to four-atomic molecules as shown by Iachello et al., by coupling three $SO(4)$ algebras, in a series of papers analyzing for example spectroscopic properties of the molecules C_2H_2, C_2D_2, C_2HD, and HCCF [18–20]. Note that, here the coupling of the first two $SO(4)$ algebras will give (τ_1, τ_2) irreps and then these are coupled to the $(\omega_3, 0)$ irreps of the third $SO(4)$ algebra. As $SO(4) \sim SO(3) \otimes SO(3)$, the algebra here also is carried out by exploiting angular momentum algebra.

12.4 Coupled *U(2)* Algebras for Vibrational Modes in Polyatomic Molecules

Study of the vibrational excited states in medium and large molecules is an important current area of research. Based on the fact that $U(2) \supset SU(2) \supset SO(2)$ [with boson number N denoting $U(2)$ irreps, $\frac{N}{2}$

the irreps of $SU(2)$ and $\frac{N}{2} - v$ the irreps of $SO(2)$] is the algebra of one-dimensional Morse oscillator, a coupled $U(2)$ model for vibrational states in polyatomic molecules has been introduced by Iachello and Oss [5] by attaching a $U_i(2)$ algebra to each bond of a polyatomic molecule. Then, the SGA for stretching vibrations is $\sum_i U_i(2)\oplus$. The interaction between any two bonds i and j is then generated by (I) local $U_i(2) \oplus U_j(2) \supset SO_i(2) \oplus SO_j(2) \supset SO_{ij}(2)$ algebra and (II) normal $U_i(2) \oplus U_j(2) \supset U_{ij}(2) \supset SO_{ij}(2)$ algebra. Note that for simplicity, the $SU(2)$ is dropped everywhere but one need to remember that $SU(2) \supset SO(2)$ algebra is the simple angular momentum algebra with the J quantum number being $\frac{N}{2}$ and the J_z quantum number being $m = \frac{N}{2} - v$. Then, the local basis is $|N_i, v_i, N_j, v_j, \rangle$. Each bond energy is generated by $C_i = [2J_z(i)]^2 - N_i^2$ with

$$\langle C_i \rangle^{N_i,v_i} = -4(N_i v_i - v_i^2). \tag{12.6}$$

It is important to note that the one-dimensional Morse oscillator is given by $h_m = p^2/2\mu + D\left[1 - \exp(-ax)\right]^2 = a_0 + a_1 C$. Similarly, the pair energy operator preserving (I) is $C_{ij} = [2J_z(i) + 2J_z(j)]^2 - (N_i + N_j)^2$ and its matrix elements are:

$$\langle C_{ij} \rangle^{N_i,v_i,N_j,v_j} = -4\left[(N_i + N_j)(v_i + v_j) - (v_i + v_j)^2\right]. \tag{12.7}$$

The interaction between the bonds i and j will mix the local (I) basis states. A simple operator for this purpose is the Majorana operator M_{ij} that is related to the Casimir operator of $SU_{ij}(2)$. The M_{ij} operator and its matrix elements (they will follow easily from the angular momentum algebra):

$$M_{ij} = -\left\{2\left[J_z(i)J_z(j) - \frac{N_i}{2}\frac{N_j}{2}\right] + J_+(i)J_-(j) + J_-(i)J_+(j)\right\},$$

$$\langle N_i v_i N_j v_j \mid M_{ij} \mid N_i v_i N_j v_j \rangle = (N_i v_j + N_j v_i - 2v_i v_j),$$

$$\langle N_i v_i - 1 N_j v_j + 1 \mid M_{ij} \mid N_i v_i N_j v_j \rangle$$
$$= -\sqrt{(N_j - v_j)(N_i - v_i + 1)v_i(v_j + 1)},$$

$$\langle N_i v_i + 1 N_j v_j - 1 \mid M_{ij} \mid N_i v_i N_j v_j \rangle$$
$$= -\sqrt{(N_i - v_i)(N_j - v_j + 1)v_j(v_i + 1)}. \tag{12.8}$$

Now, diagonalizing the following Hamiltonian:

$$H = E_0 + \sum_i^n A_i C_i + \sum_{i<j}^n A'_{ij} C_{ij} + \sum_{i<j}^n \lambda_{ij} M_{ij} \qquad (12.9)$$

in the local basis $\prod_i |N_i v_i\rangle$ will give the vibrational energies. However, molecules carry point group symmetries (ex: octahedral O_h for XY_6, D_{6h} for C_6H_6) and they need to be incorporated in Equation (12.9). It is recognized that this can be done easily by imposing restrictions on the parameters A, A' and more importantly on λ_{ij}.

Let us consider the Benzene molecule C_6H_6 as shown in Figure 12.1. There are six bonds and they are all equal imposing the conditions $N_i = N$, $A_i = A$ and $A'_{ij} = A'$. The $\sum_{i<j}^6 \lambda_{ij} M_{ij}$ term is constrained by D_{6h} symmetry depending on (i, j) nearest neighbors, next nearest neighbors and so on. Simplest choice is $S = \sum_{i<j}^6 \lambda_{ij} M_{ij}$ with $\lambda_{ij} = 1$. Next is $S' = \sum_{i<j}^6 \lambda_{ij} M_{ij}$ with $\lambda_{ij} = 1$ for nearest neighbors and zero otherwise. The nearest neighbors are with $(ij) = (12)$, (16), (23), (34), (45) and (56). Third choice is $S'' = \sum_{i<j}^6 \lambda_{ij} M_{ij}$ with $\lambda_{ij} = 1$ for next nearest neighbors and zero otherwise. The next nearest neighbors are with $(ij) = (13)$, (15), (24), (26), (35) and (46). With these, the H that generates states with D_{6h} symmetry is $H = E_0 + AC + A'C' + \lambda S + \lambda'S' + \lambda''S''$ where $C = \sum C_i$ and $C' = \sum_{i<j} C_{ij}$. Instead of constructing the H and diagonalizing it in the local basis, it is also possible to directly construct H in the symmetry adopted basis [21]. The algebraic method is applied successfully to stretching

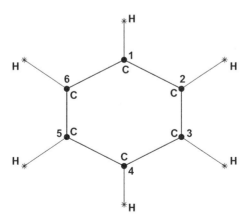

Figure 12.1 Schematic figure showing benzene molecule (C_6H_6) with D_{6h} symmetry.

overtones of SF_6, WF_6 and UF_6 molecules, C–H stretching and C–H bending vibration levels in C_6H_6 (also C_6D_6), CH stretches in n-alkane molecules and so on, see [5, 6, 22] and references therein.

12.5 $U(3)$ Algebra Chains for Bending Vibrations and Coupled Benders

In Sections 12.2 and 12.3, the full three-dimensional $U(4)$ and coupled $U(4)$ algebras for diatomic to polyatomic molecules are briefly described and in Section 12.4 coupled one-dimensional $U(2)$ algebras for stretching vibrations are described. However, even if one separates rotations and vibrations, one-dimensional description will not suffice for bending vibrations as these require two dimensions, say x and y. Introducing boson creation operators $\tau_x^\dagger$ and $\tau_y^\dagger$ together with a scalar boson creation operator $\sigma^\dagger$, we have three boson creation (call them $b_i^\dagger$, $i = 1$, 2 and 3, respectively) and three annihilation (b_i, $i = 1$, 2 and 3) operators. Then, clearly the SGA is $U(3)$ generated by the 9 operators $b_i^\dagger b_j$, $i, j = 1, 2, 3$. In order to find the subalgebras in $U(3)$, it is more convenient to consider circular bosons, $\tau_\pm^\dagger = 1/\sqrt{2}\left(\tau_x^\dagger \pm i\tau_y^\dagger\right)$ and $\tau_\pm = 1/\sqrt{2}\left(\tau_x \mp i\tau_y\right)$ satisfying the commutation relations $\left[\tau_i^\dagger, \tau_j^\dagger\right]_- = 0$, $[\tau_i, \tau_j]_- = 0$ and $\left[\tau_i, \tau_j^\dagger\right]_- = \delta_{ij}$. With these, the number operator $\hat{n}$ giving number (n) of circular bosons is $\hat{n} = \tau_x^\dagger \tau_x + \tau_y^\dagger \tau_y = \tau_+^\dagger \tau_+ + \tau_-^\dagger \tau_- = \hat{n}_+ + \hat{n}_-$. Similarly, $\hat{n}_s = \sigma^\dagger \sigma$ gives number of scalar bosons. The total boson number $N = n + n_s$ is generated by $\hat{N} = \hat{n} + \hat{n}_s$. Given N bosons, it is easy to recognize that all the N boson states belong to the totally symmetric irrep $\{N\}$ of $U(3)$. Also, it is well known that $U(3)$ admits two subalgebras [12]: (I) $U(3) \supset SO(3) \supset SO(2)$; (II) $U(3) \supset [U(2) \supset SU(2) \supset SO'(2)] \otimes U(1)$. Let us now identify the generators of the various algebras in (I) and (II) and the associated irrep reductions [7, 8].

Starting with (I), the $SO(3)$ algebra is generated by the three operators $D_+ = \sqrt{2}\left(\tau_+^\dagger \sigma - \tau_- \sigma^\dagger\right)$, $D_- = \sqrt{2}\left(\tau_+ \sigma^\dagger - \tau_-^\dagger \sigma\right)$ and $D_0 = \hat{\ell} = \hat{n}_+ - \hat{n}_-$. This is established by proving $[D_+, D_-]_- = 2D_0$ and $[D_0, D_+]_- = D_+$. The associated angular momentum quantum number is denoted by ω and the eigenvalues of:

$$\hat{W}^2 = D_+ D_- + \hat{\ell}^2 - \hat{\ell} = \frac{1}{2}\left(D_+ D_- + D_- D_+\right) + \hat{\ell}^2 \qquad (12.10)$$

are $\omega(\omega + 1)$. Then, the symmetry limit I is:

$$
\left|
\begin{array}{ccc}
U(3) & \supset & SO(3) & \supset & SO(2) \\
\{N\} & & \omega & & \ell \\
& & \left(D_+, D_-, \hat{\ell}\right) & & \hat{\ell}
\end{array}
\right\rangle
\tag{12.11}
$$

$$N \to \omega = N, N - 2, \ldots, 0 \text{ or } 1$$
$$\omega \to \ell = -\omega, -\omega + 1, \ldots, 0, \ldots, \omega - 1, \omega.$$

As discussed ahead, ω quantum number is also related to pairing. As in Section 12.2, introducing the vibrational quantum number $v = (N - \omega)/2$ will give $v = 0, 1, 2, \ldots, (N/2)$ or $(N-1)/2$ and $\ell = 0, \pm 1, \pm 2, \ldots, \pm(N - 2v)$. Now, the basis states are $|N, v, \ell\rangle$ and a Hamiltonian (including at most quadratic Casimir invariants C_r, $r \leq 2$) preserving the symmetry limit I is $H = E_0 + \alpha C_1(U(3)) + A C_2(SO(3)) + B\left[C_1(SO(2))\right]^2$ giving $E = E_0 + \alpha N + AN(N + 1) - 4A\left[(N + \frac{1}{2})v - v^2\right] + B\ell^2$.

In $U(N) \supset SO(N)$ for bosons, the $SO(N)$ is related to pairing [23]. This result applies to $SO(3)$ in $U(3) \supset SO(3)$. With $\tau_\pm$ and σ bosons, the pair creation operator $\hat{P} = 2\tau_+^\dagger \tau_-^\dagger + \sigma^\dagger \sigma^\dagger$. Then, the pairing Hamiltonian, a two-body operator, is $H_p = \hat{P}(\hat{P})^\dagger$; note that $(\hat{P})^\dagger = (2\tau_+\tau_- + \sigma\sigma)$. Simple algebra gives the important relation:

$$H_p = \hat{P}(\hat{P})^\dagger = \hat{N}(\hat{N} + 1) - (\hat{W})^2 \tag{12.12}$$

establishing the relation between pairing and the $SO(3)$ algebra. It is also important to point out that there is a second $SO(3)$ subalgebra in $U(3)$ and we will denote this by $\overline{SO(3)}$. Its generators and the corresponding pairing operator H'_P are:

$$
\overline{SO(3)} \leftrightarrow (R_+, R_-, \hat{\ell}),
$$
$$
R_+ = \sqrt{2}(\tau_+^\dagger \sigma + \tau_- \sigma^\dagger), \quad R_- = \sqrt{2}(\tau_-^\dagger \sigma + \tau_+ \sigma^\dagger),
$$
$$
[R_+, R_-] = 2\hat{\ell}, \quad [\hat{\ell}, R_+] = R_+,
\tag{12.13}
$$
$$
\hat{R}^2 = R_+ R_- + (\hat{\ell})^2 - \hat{\ell} \to \left\langle \hat{R}^2 \right\rangle^{N,\omega,\ell} = \omega(\omega + 1),
$$
$$
P' = 2\tau_+^\dagger \tau_-^\dagger - \sigma^\dagger \sigma^\dagger, \quad H'_P = P'(P')^\dagger = \hat{N}(\hat{N} + 1) - \hat{R}^2.
$$

For the significance of $U(3) \supset \overline{SO(3)} \supset SO(2)$ see Appendix.

Turning to limit II, it is easy to recognize that we can divide the space into the one with τ bosons and other with σ bosons giving $U(3) \supset U_\tau(2) \oplus U_\sigma(1)$

with $U(2)$ generating n and $U(1)$ generating n_s so that $N = n + n_s$. As we always consider states with a fixed N value, given n the value of n_s is uniquely $N - n$ and therefore we will not mention $U(1)$ hereafter. The $U(2)$ algebra is generated by the 4 operators $Q_+ = \tau_+^\dagger \tau_-$, $Q_- = \tau_-^\dagger \tau_+$, $Q_0 = (\hat{n}_+ - \hat{n}_-)/2 = \hat{\ell}/2$ and $\hat{n}$. More importantly, the operators $\{Q_+, Q_-, Q_0\}$ form angular momentum algebra $SU(2)$ with m quantum number $\ell/2$. Given n bosons, the $SU(2)$ irrep is spin $\frac{n}{2}$ with $\ell/2$ taking values from $-\frac{n}{2}$ to $\frac{n}{2}$. Putting all these together, the symmetry limit II is:

$$
\left| \begin{array}{ccccc}
U(3) & \supset & U(2) & \supset & SO(2) \\
\{N\} & & n & & \ell \\
& & \left(Q_+, Q_-, \hat{\ell}/2, \hat{n} \right) & & \hat{\ell}
\end{array} \right\rangle
\tag{12.14}
$$

$$
N \rightarrow n = N, N - 1, \ldots, 0,
$$
$$
n \rightarrow \ell = \pm n, \pm(n - 2), \ldots, 0 \text{ or } 1.
$$

Now, the basis states are $|N, n, \ell\rangle$ and a Hamiltonian (including at most quadratic Casimir invariants C_r, $r \le 2$) preserving the symmetry limit II is $H = E_0 + \alpha C_1(U(2)) + \beta C_2(U(2))) + B\left[C_1(SO(2))\right]^2$ giving $E = E_0 + \alpha n + \beta n(n+1) + B\ell^2$.

Most general $U(3)$ Hamiltonian preserving N and ℓ can be written as a polynomial in the nine $U(3)$ generators $\hat{N}$, $\hat{n}$, $\hat{\ell}$, $D_\pm$, $R_\pm$ and $Q_\pm$. Note that D and R operators change ℓ by one unit and Q by two units. It is easy to write the matrix elements of H (i.e. construct H matrix) in the $|N, n, \ell\rangle$ basis; $n_s = N - n$, $n_+ = (n + \ell)/2$ and $n_- = (n - \ell)/2$. Both second degree and higher degree polynomials are used in the applications to bending motion in many triatomic molecules [8, 24]. Another important aspect of the $U(3)$ model is that the simple interpolating Hamiltonian:

$$
H_{mix} = (1 - \xi)\,\hat{n} + \frac{\xi}{N - 1} H_P
\tag{12.15}
$$

captures the essence of the two limits I and II. Note that H_P is defined before and its matrix elements in the $|N, n, \ell\rangle$ basis follow easily from its definition,

$$
\langle N, n', \ell \mid H_P \mid N, n, \ell\rangle = \left[(N - n)(N - n - 1) + n^2 - \ell^2\right]\delta_{n',n}
$$
$$
+ \sqrt{(N - n + 2)(N - n + 1)(n + \ell)(n - \ell)}\,\delta_{n',n-2}
$$
$$
+ \sqrt{(N - n)(N - n - 1)(n + \ell + 2)(n - \ell + 2)}\,\delta_{n',n+2}.
\tag{12.16}
$$

As shown in [8], with ξ varying from 0 to 1 the Hamiltonian changes the structure from rigidly linear ($\xi = 0$) to rigidly bent ($\xi = 1$) structure. More importantly, for $0 < \xi \leq 0.2$, the molecule will be quasi-linear and for $0.2 < \xi < 1$ quasi-bent. Moreover, at $\xi = 0.2$ the system exhibits QPT (change in ground state structure) and it is a second order phase transition. Also, at $\xi = 0.6$ the system with H defined by Equation (12.15) exhibits excited state quantum phase transition (EQPT) [24]. Let us stress that the $U(3)$ model is a simple two-level model ($3 = 2 + 1$) and the QPT and EQPT are typical of general two-level models [25, 26]. For bosons in two-levels with n_1 and n_2 number of degenerate single-particle levels, the SGA is $U(n_1 + n_2)$ and then there are two symmetry limits, S1: $U(n_1 + n_2) \supset U(n_1) \oplus U(n_2) \supset SO(n_1) \oplus SO(n_2) \supset K$ and S2: $U(n_1 + n_2) \supset SO(n_1 + n_2) \supset SO(n_1) \oplus SO(n_2) \supset K$. In generating the spectrum for a fixed $SO(n_1) \oplus SO(n_2)$ irrep, the Lie algebra K will not play any role. Numerical examples for QPT (also EQPT) are shown in Figures 12.2 and 12.3 for some general two-level models, see Refs. [27–30] for details of the results in the figures.

Besides describing single benders, using coupled $U(3)$ algebras it is possible to study various structures generated by coupled benders in tetra-atomic

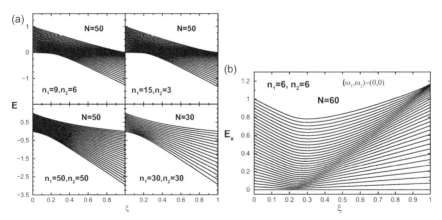

Figure 12.2 (a) Spectra as a function of the mixing parameter ξ in H interpolating the symmetry limits (S1) and (S2) mentioned in Section 12.5. Results are shown for: (i) (n_1, n_2) = (9,6) and (15,3) with number of bosons N = 50; (ii) (n_1, n_2) = (50,50) with N = 50; (iii) (n_1, n_2) = (30,30) with N = 30. All results are shown for (ω_1, ω_2) = (0,0) irrep where ω_1 is the irrep of $SO(n_1)$ and ω_2 is the irrep of $SO(n_2)$. (b) Excitation energies as a function of the mixing parameter ξ for (n_1, n_2) = (6,6), N = 60 and (ω_1, ω_2) = (0,0). Figures show that there will be QPT only when the boson number N is much greater than $n_1 + n_2$. Figure is constructed from the results in Ref. [27] and see this reference for further details.

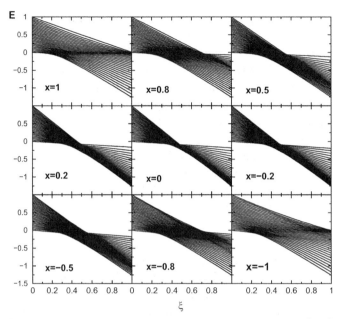

Figure 12.3 Energy spectra for 50 bosons with the Hamiltonian $H(\xi, x) = (1 - \xi)/N^B\, \hat{n}_2 + \left[(\xi/(N^B)^2) \right] \left[4(S_+^1 + xS_+^2)(S_-^1 + xS_-^2) - N^B(N^B + 13) \right]$ interpolating the symmetry limits (S1) with $U(15) \supset U(6) \oplus U(9) \supset SO(6) \oplus SO(9)$ and (S2) with $U(15) \supset SO(15) \supset SO(6) \oplus SO(9)$ mentioned in Section 12.5 with two orbits having degeneracies $n_1 = 6$ and $n_2 = 9$, respectively. Note that S_+^1 is the pair creation operator for the first orbit and S_+^2 for the second orbit. Similarly, N^B is total boson number operator and $\hat{n}_2$ boson number operator for the second orbit. All the results are for $(\omega_1, \omega_2) = (0, 0)$ where ω_1 is the $SO(6)$ irrep and ω_2 is the $SO(9)$ irrep. In each panel, energy spectra are shown as a function of the parameter ξ taking values from 0 to 1. Results are shown in the figures for $x = 1, 0.8, 0.5, 0.2, 0, -0.2, -0.5, -0.8$ and -1; $x = 1$ and -1 correspond to the two $SU(1, 1)$ algebras in the model. In the figures, energies are not in any units, see Ref. [30] for further details.

molecules. Associating a $U(3)$ for each bender, we have $U_1(3) \oplus U_2(3)$ SGA with large number of subalgebra chains preserving boson numbers N_1 and N_2 and the total $\ell = \ell_1 + \ell_2$ quantum number. At the first level, the subalgebras are $U_1(2) \oplus U_2(2)$, $SO_1(3) \oplus SO_2(3)$, $U_{12}(3)$ and $U_1(2) \oplus SO_2(3)$. The $U_1(2) \oplus U_2(2)$ admits $U_{12}(2)$ and $SO_1(2) \oplus SO_2(2)$ subalgebras, $SO_1(3) \oplus SO_2(3)$ admits $SO_{12}(3)$ and $SO_1(2) \oplus SO_2(2)$ subalgebra [the later also appears in $U_1(2) \oplus SO_2(3)$] and finally $U_{12}(3)$ admits $U_{12}(2)$ and $SO_{12}(3)$ subalgebras. All these will have the final subalgebra $SO_{12}(2)$. It is possible to write the generators of all these algebras and also one- and two-body operators that preserve N and ℓ. Extending the algebras described before for one bender, it is possible to construct the H matrix for the coupled benders systems. However, a simple Hamiltonian describing the various structures is of the form $H = H_1 + H_2 + V_{12}$ with H_i same as discussed above for one bender and V_{12} contains, $(P_{12})(P_{12})^\dagger$, $\hat{W}_1 \cdot \hat{W}_2$, the quadratic Casimir invariant of $SU_{12}(3)$ or equivalently the Majorana operator M_{12} and so on. See [9, 31] for further mathematical details and applications to C_2H_2 and H_2CO molecules.

12.6 Conclusions and Future Outlook

Starting with the $U(4) \supset SO(4) \supset SO(3)$ Lie algebra chain for rotation–vibration levels in diatomic molecules, a brief account of the Lie algebraic approach to triatomic molecules using two coupled $SO(4)$ algebras, for four-atomic molecules three coupled algebras, coupled $SU(2)$ algebras for polyatomic molecules, and $U(3)$ algebra for coupled benders is given in Sections 12.2–12.5. The Lie algebra approach is not too complex, and yet, it is powerful as seen from the applications carried out till now. In this short review, all mathematical details are kept to a minimum, and for detailed comparisons between theory and experimental data, the references given at the end should be consulted. There are several new directions enlarging the scope of the Lie algebra approach, and some of these are as follows: (i) Extension of the analysis of two coupled benders (in four-atomic molecules) using $U(3)$ algebra to three or more benders. (ii) Algebraic approach for simultaneous description of electronic, vibrational, and rotational energy levels. For example, with electrons in s and p orbitals, the SGA for electrons is $U(8) \supset U(4) \otimes SU(2)$ with $U(4)$ for the spatial part and $SU(2)$ generating spin. The key point now is that the spatial $U(4)$ can be combined with the $U(4)$ generated by (π, s) bosons to give a Bose-Fermi (BF) coupling scheme [13]. Let us add that BF schemes are well studied in nuclear structure [10, 11]. (iii) Development of the Lie algebra approach and its applications to polyatomic molecules with very large number of atoms (also to macromolecules, polymers, etc.) [5, 22, 32]. (iv) Shape phase transitions that correspond to QPT and EQPT can be studied using classical analysis of the Lie algebraic models and with this it is possible to address quantum monodromy in molecules [8, 24, 25]. In fact quantum monodromy is seen recently in some molecules [33]. Some aspects of QPT and EQPT are mentioned in Section 12.5. (v) Applications of the algebraic coupling schemes discussed in Sections 12.2–12.5 in the study of order chaos transitions and random matrix theory, see Appendix and Ref. [34]. (vi) Level densities, partition functions, and other thermodynamic quantities can be studied using the algebraic models; some analytical results for diatomic and triatomic molecules are available as presented in the Appendix.

Appendix

Symmetry Mixing Hamiltonians Generating Regular Spectra

Given the two symmetry limits (i) $U(4) \supset SO(4) \supset SO(3)$ and (ii) $U(4) \supset [SU(3) \supset SO(3)] \oplus U(1)$ for diatomic molecules, general two-body Hamiltonian mixing these two symmetry limits is:

$$H_{mix} = \alpha_0(N) + \alpha_1 C_1(U(1)) + \alpha_2 C_2(SO(4))$$
$$+ \alpha_3 C_2(SU(3)) + \alpha_4 C_2(SO(3)). \tag{12.17}$$

Note that $\alpha_0(N)$ is a quadratic polynomial in N. More importantly, $\alpha_2 = 0$ will give limit (ii) and $\alpha_1 = \alpha_3 = 0$ will give limit (i). However, even when $\alpha_1, \alpha_2, \alpha_3 \neq 0$, it is possible to produce a regular spectrum. This is due to the existence of $\overline{SO(4)}$ generated by L_q^1 and $\mathcal{D}_\mu^1 = (\pi^\dagger s - s^\dagger \tilde{\pi})_\mu^1$ mentioned in Section 12.2. Note that both $C_2(SO(4))$ and $C_2(\overline{SO(4)})$ generate the same spectrum with eigenvalues $\omega(\omega + 2)$. These operators are given by:

$$C_2(SO(4)) = 2\left(\pi^\dagger\tilde{\pi}\right)^1 \cdot \left(\pi^\dagger\tilde{\pi}\right)^1 - (\pi^\dagger s + s^\dagger\tilde{\pi})^1 \cdot (\pi^\dagger s + s^\dagger\tilde{\pi})^1$$
$$C_2(\overline{SO(4)}) = 2\left(\pi^\dagger\tilde{\pi}\right)^1 \cdot \left(\pi^\dagger\tilde{\pi}\right)^1 + (\pi^\dagger s - s^\dagger\tilde{\pi})^1 \cdot (\pi^\dagger s - s^\dagger\tilde{\pi})^1$$
$$\Rightarrow C_2(SO(4)) + C_2(\overline{SO(4)}) = 4\left(\pi^\dagger\tilde{\pi}\right)^1 \cdot \left(\pi^\dagger\tilde{\pi}\right)^1$$
$$-2\left[\pi^\dagger s \cdot s^\dagger\tilde{\pi} + s^\dagger\tilde{\pi} \cdot \pi^\dagger s\right]. \tag{12.18}$$

Using the results that $ss^\dagger = (n_s + 1)$ and $\tilde{\pi} \cdot \pi^\dagger = -(3 + n_\pi)$ we have:

$$C_2(\overline{SO(4)}) = -C_2(SO(4)) + 4(N-1)n_\pi - 4n_\pi^2 + 6N + 2C_2(SO(3)). \tag{12.19}$$

As $C_1(U(1)) = n_s = N - n_\pi$ and $C_2(SU(3)) = n_\pi(n_\pi + 3)$, clearly for a particular choice of the parameters in Equation (12.17), H_{mix} can be reduced to $C_2(\overline{SO(4)})$ and hence solvable (generates a regular spectrum). For details of the significance of this result for order-chaos transitions and quantum phase transitions, see Refs. [25, 35]. It is also important to add that the occurrence of multiple pairing algebras, as seen from the $U(3)$ model discussed in Section 12.5, is a general feature of both fermion and boson systems with two or more levels and they play an important role in QPT and EQPT, see Refs. [29, 30].

Partition Functions for Diatomic and Triatomic Molecules

Starting with the energy formula given by Equation (12.2), it is possible to derive a simple formula for the partition function $Z(\beta) = Tr(\exp -\beta E)$ for diatomic molecules in the $SO(4)$ [$U(4) \supset SO(4) \supset SO(3) \supset SO(2)$] limit. Using Equation (12.2) for the eigenvalues and the allowed quantum numbers, we have:

$$Z_{SO(4)}(\beta) = Z_0 \sum_{v=0}^{N/2} \sum_{L=0}^{N-2v} (2L+1) \exp -\beta\{A(N+1)v - Av^2 + BL(L+1)\}.$$

(12.20)

Note that $A = -4\alpha$ and $B = \beta$; α and β are defined in Equation (12.2). In addition, Z_0 is a constant. Clearly, with $A > 0$ and $B > 0$ the ground state is $|N, v = 0, L = 0\rangle$. With $N \to \infty$ and $\sigma = 1/(2\beta B)^{1/2} \gg 1$, $Z(\beta)$ takes the simpler form, with $\int_0^\infty (2L + 1) \exp -L(L+1)/2\sigma^2 \, dL = 2\sigma^2$,

$$Z_{SO(4)}(\beta) \stackrel{N\to\infty, \sigma\gg 1}{\longrightarrow} Z_0 \left(2\sigma^2\right) \sum_{v=0}^{\infty} \exp -\beta(AN)v$$

$$= Z_0 \left(2\sigma^2\right) \left(1 - \exp -\beta(AN)\right)^{-1} = Z_0 \, Z_{rot}(\beta) \, Z_{vib}(\beta).$$

(12.21)

Note that $Z_{rot}(\beta) = 2\sigma^2$, $Z_{vib}(\beta) = (1 - \exp -\beta(AN))^{-1}$ and $\sigma^2 = 1/(2\beta B)$. The decomposition of $Z_{SO(4)}(\beta)$ into a product of Z's for the rotational and vibrational parts is similar to the decomposition obtained before in the interacting boson model of atomic nuclei [36]. A different formula, in the limit $\beta \to 0$, is given by:

$$Z_{SO(4)}(\beta) \stackrel{\beta\to 0}{\longrightarrow} Z_0 \int_0^{N/2} dv \, f(v) \int_0^{N-2v} dL \, (2L+1) \exp -\beta BL(L+1)$$

$$= Z_0 \frac{1}{B\beta} \int_0^{N/2} dv \, \{1 - \exp -\beta B(N-2v)(N-2v+1)\} \, f(v);$$

$$f(v) = \exp -\beta \left\{A(N+1)v - Av^2\right\}.$$

(12.22)

The last integral here can be written in terms of error functions; see also Ref. [37]. Let us add that more accurate formulas for $Z_{SO(4)}(\beta)$ can be derived using Euler-Maclaurin summation formula.

The other symmetry limit, for diatomic molecules, starting with $U(4)$ is $U(4) \supset [SU(3) \supset SO(3) \supset SO(2)] \oplus U(1)$ with basis states $|N, n_\pi, L, M\rangle$

where $N \to n_\pi = 0, 1, \ldots, N$ and $n_\pi \to L = n_\pi, n_\pi - 2, \ldots, 0$ or 1. Now, the Hamiltonian and the partition function in the $SU(3)$ limit are:

$$H = E_0'' + A_1 C_1(U(3)) + A_2 C_2(SU(3)) + A_3 L(L+1),$$

$$Z_{SU(3)}(\beta) = Z_0 \sum_{n_\pi=0}^{N} \sum_{L \in n_\pi} (2L+1)$$
$$\times \exp -\beta \{A_1 n_\pi + A_2 n_\pi (n_\pi + 3) + A_3 L(L+1)\}.$$
$$(12.23)$$

Note that with $A_1 > 0$, $A_2 << A_3$ and $A_3 > 0$, the ground state is $|N, n_\pi = 0, L = 0\rangle$. In the symmetry limit it is a good approximation to assume $A_2, A_3 \simeq 0$. Then we have:

$$Z_{SU(3)}(\beta) = Z_0 \sum_{n_\pi=0}^{N} \frac{(n_\pi + 1)(n_\pi + 2)}{2} \exp -\beta(A_1 n_\pi) \qquad (12.24)$$
$$= Z_0 (1 - \exp -\beta A_1)^{-2}.$$

In addition, it is also possible to derive a formula for $Z_{SU(3)}(\beta)$ in the $\beta \to 0$ limit in terms of error functions.

Turning to tri-atomic molecules, using the energy formula given by Equation (12.5) and the associated quantum numbers (see Section 12.3), it is possible to derive a formula for the partition function $Z_{local-SO_{12}(4)}(\beta)$ in the local basis symmetry limit $U_1(4) \oplus U_2(4) \supset SO_1(4) \oplus SO_2(4) \supset SO_{12}(4) \supset SO(3)$. In the limit $N_1 \to \infty$, $N_2 \to \infty$, $a_{12} \sim 0$ and $\sigma >> 1$ (a_{12} is the strength of $C_2(SO_{12}(4))$ and $\sigma^2 = 1/2\beta d$ where d is the strength of $L(L+1)$ term), the energy formula given by Equation (12.5) reduces to the form $E(N_1, N_2, v_1, v_2^{\ell_2}, v_3, L) = e_0 + e_1 v_1 + e_2 v_2 + e_3 v_3 + dL(L+1)$. Then $Z(\beta)$ is:

$$Z_{local-SO_{12}(4)}(\beta) = Z_0 \sum_{v_1,v_2,v_3=0}^{\infty} \sum_{\ell_2 \in v_2} \sum_{L} (2L+1) \exp$$
$$- \beta \{e_1 v_1 + e_2 v_2 + e_3 v_3 + dL(L+1)\}. \quad (12.25)$$

The L integration gives $2\sigma^2$ for $\ell_2 = 0$ and $2(2\sigma^2)$ for $\ell_2 \neq 0$; see Equation (12.4) for $\tau_2 = \ell_2 \to L$ and the doubling for $\ell_2 \neq 0$. Combining this with the $v_2 \to \ell_2$ reductions ($v_2 = 0 \to \ell_2 = 0$, $v_2 = 1 \to \ell_2 = 1$, $v_2 = 2 \to \ell_2 = 0, 2$, $v_2 = 3 \to \ell_2 = 1, 3$, $v_2 = 4 \to \ell_2 = 0, 2, 4, \ldots$) will allow us to carry out the ℓ_2 summation in Equation (12.25) giving:

$$
\begin{aligned}
Z_{local-SO_{12}(4)}(\beta) \\
= (2\sigma^2) \sum_{v_1,v_2,v_3=0}^{\infty} (v_2+1)\, \exp-\beta(e_1 v_1 + e_2 v_2 + e_3 v_3) \\
= (2\sigma^2)\,(1-\exp-\beta e_1)^{-1}\,(1-\exp-\beta e_3)^{-1}\,(1-\exp-\beta e_2)^{-2}.
\end{aligned}
\tag{12.26}
$$

Further improvements of the formula for $Z_{local-SO_{12}(4)}(\beta)$ are possible. Also, in future it is important to derive the formulas for $Z(\beta)$ for the other symmetry limits of the $U_1(4) \oplus U_2(4)$ model.

Acknowledgement

Thanks are due to F. Iachello for many useful discussions during several visits to Yale.

References

[1] F. Iachello and A. Arima. *The Interacting Boson Model*. Cambridge University Press, 1987.

[2] V. K. B. Kota and R. Sahu. *Structure of medium mass nuclei: deformed shell model and spin-isospin interacting boson model*. CRC Press of Taylor & Francis, 2017.

[3] F. Iachello and R. D. Levine. *Algebraic Theory of Molecules*. Oxford University Press, 1995.

[4] A. Frank and P. Van Isacker. *Algebraic Methods in Molecular and Nuclear Physics*. John Wiley & Sons, 1994.

[5] F. Iachello and S. Oss. Model of n coupled anharmonic oscillators and applications to octahedral molecules. *Physical Review Letters*, 66: 2976–2979, 1991; Algebraic methods in quantum mechanics: From molecules to polymers. *Euro Physics Journal D*, 19:307–314, 2002.

[6] K. S. Rao, J. Choudhury, N. K. Sarkar and R. Bhattacharjee. Vibrational spectra of nickel metalloporphyrins: An algebraic approach. *Pramana – Journal of Physics*, 72:517–525, 2009; K. S. Rao, V. U. M. Rao and J. Vijayasekhar. Vibrational Spectra of Polyatomic Molecules Using Lie Algebraic Method: (A Review). *Oriental Journal of Chemistry*, 32(1):437–440, 2016.

[7] F. Iachello and S. Oss. Algebraic approach to molecular spectra: Two dimensional problems. *Journal of Chemical Physics*, 104:6956–6963, 1996.

[8] F. Perez-Bernal and F. Iachello. Algebraic approach to two-dimensional systems: Shape phase transitions, monodromy, and thermodynamic quantities. *Physical Review A*, 77:032115/1–21, 2008.

[9] F. Iachello and F. Perez-Bernal. A novel algebraic scheme for describing coupled benders in tetra-atomic molecules. *Journal of Physical Chemistry A*, 113:13273–13286, 2009.

[10] F. Iachello and P. Van Isacker. *The interacting boson – fermion model*. Cambridge University Press, 1991.

[11] R. Bijker and V. K. B. Kota. Interacting boson – fermion model of collective states: boson – fermion symmetries related to the U(5) limit. *Annals of Physics (New York)*, 156:110–141, 1984; Interacting boson fermion model of collective states: The $SU(3) \otimes U(2)$ limit. *Annals of Physics (New York)*, 187:148–197, 1988.

[12] V. K. B. Kota and Y. D. Devi. *Nuclear shell model and the interacting boson model: Lecture notes for practitioners*. Published by IUC-DAEF Calcutta Center, Kolkata, 1996.

[13] A. Frank, R. Lemus and F. Iachello. Algebraic approach to molecular electronic spectra I. Energy levels. *Journal of Chemical Physics*, 91:29–41, 1989.

[14] V. K. B. Kota. Group Theoretical and Statistical Properties of Interacting Boson Models of Atomic Nuclei: Recent Developments. In A.V. Ling, editor, *Focus on Boson Research*, pp. 57–105. Nova Science Publishers Inc., New York, 2006.

[15] J. L. Dunham. The energy levels of a rotating vibrator. *Physical Review* 41:721–731, 1932.

[16] F. Iachello and S. Oss. Overtone frequencies and intensities of bent XY_2 molecules in the vibron model. *Journal of Molecular Spectroscopy*, 142:85–107, 1990; Vibrational spectra of linear triatomic molecules in the vibron model. *Journal of Molecular Spectroscopy*, 146:56–78, 1991.

[17] N. K. Sarkar, J. Choudhury and R. Bhattacharjee. An algebraic approach to the study of the vibrational spectra of HCN. *Molecular Physics*, 104:3051–3055, 2006.

[18] J. Hornos and F. Iachello. The overtone spectrum of acetylene in the vibron model. *Journal of Chemical Physics*, 90:5284–5293, 1989.

[19] F. Iachello, S. Oss and R. Lemus. Linear four-atomic molecules in the vibron model. *Journal of Molecular Spectroscopy*, 149:132–151, 1991.

[20] F. Iachello, S. Oss and L. Viola. Vibrational analysis of monofluoroacetylene (HCCF) in the vibron model. *Molecular Physics*, 78:545–559, 1993.

[21] J. Q. Chen, F. Iachello and J. L. Ping. The method of symmetrized bosons with applications to vibrations of octahedral molecules. *Journal of Chemical Physics*, 104:815–825, 1996.

[22] T. Marinkovic and S. Oss. Algebraic description of n-alkane molecules: First overtone of CH stretching modes. *Physical Chemistry Communications*, 6: 42–46, 2003.

[23] V. K. B. Kota. $O(36)$ Symmetry Limit of IBM-4 with good s, d and sd Boson Spin-Isospin Wigner's $SU(4) \sim O(6)$ Symmetries for N $\approx$ Z Odd-Odd Nuclei. *Annals of Physics (New York)*, 280:1–34, 2000.

[24] D. Larese, F. Perez-Bernal and F. Iachello. Signature of quantum phase transitions and excited state quantum phase transitions in the vibrational bending dynamics of triatomic molecules. *Journal of Molecular Structure*, 1051:310–327, 2013.

[25] P. Cejnar, J. Jolie and R. F. Casten. Quantum phase transitions in the shapes of atomic nuclei. *Review of Modern Physics*, 82:2155–2212, 2010.

[26] M. A. Caprio, P. Cejnar and F. Iachello, *Annals of Physics* (*New York*) 323, 1106 (2008); J. E. Garcia-Ramos, P. Perez-Fernandez and J. M. Arias. Excited-state quantum phase transitions in a two-fluid Lipkin model. *Physical Review C*, 95:054326/1–15, 2017.

[27] V. K. B. Kota. Interacting Boson Model Applications To Exotic Nuclear Structure. *AIP Conference Proceedings*, 1524:52–57, 2013.

[28] V. K. B. Kota. Lie Algebra Symmetries and Quantum Phase Transitions in Nuclei. *Pramana-Journal of Physics*, 82:743–755, 2014.

[29] V. K. B. Kota. Multiple multi-orbit fermionic and bosonic pairing and rotational SU(3) algebras. *Bulgarian Journal of Physics*, 44:454–465, 2017.

[30] V. K. B. Kota. Multiple multi-orbit pairing algebras in shell model and interacting boson model. arXiv:1707.03552, 2017.

[31] D. Larese, M. A. Caprio, F. Perez-Bernal and F. Iachello. A study of the bending motion in tetra-atomic molecules by the algebraic operator expansion method. *Journal of Chemical Physics*, 140:014304/1–14, 2014.

[32] F. Iachello and P. Truini. Algebraic model of anharmonic polymer chains. *Annals of Physics (New York)*, 276:120–143, 1999.

[33] M. Winnewisser et al. Pursuit of quantum monodromy in the far-infrared and mid-infrared spectra of NCNCS using synchrotron radiation. *Physical Chemistry Chemical Physics*, 16:17373–17407, 2014.

[34] V. K. B. Kota. *Embedded Random Matrix Ensembles in Quantum Physics*, Lecture Notes in Physics 884, Springer, 2014.

[35] D. Kusnezov. Incompleteness of representation theory: Hidden symmetries and quantum non-integrability. *Physical Review Letters*, 79:537–540, 1997.

[36] V. K. B. Kota. Partition functions in the Dynamical symmetry limits of the interacting boson model. *Euro Physics Letters*, 23:481–487, 1993.

[37] D. Kusnezov. Density of states for complex molecules. *Physical Review A*, 50:R2814–R2817, 1994.

13

Synthesis of Ultrafine Nanopowder of Y_2O_3-ZnO Composite by a Modified Combustion Method for Their Application as Infrared Transparent Ceramic Materials

Steffy Maria Jose, Christopher Thresiamm Mathew, Yesoda Velukutty Swapna, Jayachandran Santhakumari Lakshmi and Jijimon K. Thomas*

Department of Physics, Mar Ivanios College, Thiruvananthapuram 695015, Kerala, India
E-mail: jkthomasemrl@yahoo.com
*Corresponding Author

Infrared transparent ceramic materials are those materials separating environments of differing pressures or temperatures, while allowing energy at infrared wavelength range to pass between them. Synthesis and characterization of ultrafine nanostructured yttria-zinc oxide nanocomposites prepared by a modified combustion technique are presented in this chapter. The X-ray diffraction pattern reveals that the as-synthesized powder is composed of ZnO and Y_2O_3 phases. All the peaks of 50:50 mass % Y_2O_3-ZnO nanocomposite are indexed, and the crystallite size calculated using Scherrer formula is $\sim$16 nm for 222 peak of Y_2O_3 and 22 nm for 110 peak of ZnO. The particulate properties of the modified combustion product are studied with the help of high-resolution transmission electron microscopy. The as-prepared nanopowder is characterized by FTIR spectroscopy to ascertain its phase purity, which is an essential requirement for high-quality IR transparent material. The optical band gap of the sample determined using UV-visible spectroscopy is found to be 3.22 eV. TGA analysis confirms its thermal stability. The samples with enhanced infrared transmission properties have applications in infrared windows/domes, IR inspection ports, transparent armors, and night vision cameras. The sample is sintered at 1410°C to achieve 96.2% densification.

13.1 Introduction

Infrared transparent ceramics are those materials which isolate two environments with different temperature, pressure, humidity, and so on while allowing energy at infrared wavelength to pass through it. An infrared transparent ceramic material found applications in civilian and defense fields and is successfully used in armor configurations, IR inspection ports, infrared windows/domes, transparent armors, night vision cameras, and so on. The literature survey reveals that much work has been done to yield high-quality IR transparent ceramic materials but a pore-free nanoceramics with improved mechanical, thermal, optical properties is a major challenge in the fabrication of IR transparent ceramic materials. To fabricate high-quality infrared transparent windows, at most care should be taken throughout the process from the synthesis of ultrafine starting powder till the final lapping and polishing of the window [1]. In case of infrared transparent materials made from polycrystalline materials, there is always a trade-off between infrared transparency and mechanical strength. The optical and mechanical properties of a transparent ceramic material are highly dependent on its grain size and residual porosity. In this work, a novel and highly effective single step modified combustion technique is used to synthesis a high-quality starting powder with fine particle size. The superior quality of nanopowder thus prepared will be more suitable to yield a highly transparent infrared transparent ceramic material. Mostly, for the synthesis of IRT transparent ceramic materials, a solid-state route is used, and it requires multiple calcinations to get single phase resulting in increased particle size. Also it requires prolonged and high temperature sintering resulting in large grain growth. A large variety of IRT materials have been so far developed but the technologists have to compromise between different properties which adversely affect the quality of the IRT material. All these major challenges are overcome by the patented-modified combustion technique by synthesizing high-quality starting powder [2–6].

Even though lots of work have been done on pure yttria, the technologists have to compromise with the mechanical properties that are related to particle size growth. The addition of nanocrystalline second phases to the ordinary grain-sized ceramics is an effective way of reducing particle grain growth. Also the second-phase particles that are insoluble in the matrix phase can decrease the coarsening kinetics of the matrix by reducing the grain boundary mobility through a particle pinning effect. The ZnO second phase is added to Y_2O_3 to inhibit the matrix grain size and thus increasing the mechanical properties. The homogeneous distribution of ultrafine Y_2O_3 grain induces a pinning effect of ZnO grain boundaries which reduces the average grain size. Thus, Y_2O_3-ZnO nanocomposite is able to replace the limitations exhibited by pure yttria ceramics [7–10].

13.2 Sample Preparation and Characterization

The raw materials comprise of yttrium nitrate hexahydrate (99.9%, Alfa Aesar), zinc oxide (99.5%, HiMedia), and citric acid (99.9%, Merck). A modified combustion method is used to synthesize highly pure, homogeneous, ultrafine nanocrystalline ceramic powders of Y_2O_3-ZnO nanocomposite in a single step. Initially to prepare a clear aqueous solution of Zn and Y ions, stochiometric amount of high purity $Y(NO_3)_3 \cdot 6 H_2O$ and $Zn(NO_3)_2 \cdot 6 H_2O$ was dissolved in double distilled water. Citric acid was then added to it as a complexing agent. Amount of citric acid is calculated based on total valence of the oxidizing and the reducing agents for maximum release of energy during combustion [11]. Nitric acid was used as the oxidizing agent and ammonia as the fuel, and the pH of the solution was monitored till it becomes 7. The solution containing the precursor mixture was heated using a hot plate at 250°C in a ventilated fume hood. A precise and uniform formulation of the desired composition of the sample was formed initially from a reaction media in a liquid state. High purity and crystallinity of the sample was achieved at high reaction temperature. The solution boils on heating and undergoes dehydration to produce foam. Figure 13.1 shows the photograph of auto combustion of Y_2O_3-ZnO nanocomposite. The foam then ignites by itself on persistent heating giving a voluminous and fluffy product. The particle growth will be inhibited by various gases formed in the process leading to the formation of a voluminous and fluffy product, that is, nanosize sample powder.

The as-prepared ultrafine nanopowders obtained from the modified combustion process were characterized by different powder characterization techniques. The crystal structure of the samples was studied by X-ray diffraction (XRD) by Rigaku-Miniflex 600 with Cu Kα radiation in the range of 20–80° in a step width of 0.03° for the determination of crystalline structure and phase of the nanomaterials. The average crystallite size of all the samples was determined by Debye Scherer's equation, $D = \frac{K\lambda}{\beta cos\theta}$ [11]. The absorption spectrum of the as-prepared sample was recorded using a Shimadzu spectrophotometer (UV-1700) [12–14]. The UV–Visible spectra were recorded to resolve the correlation between the grain size and absorption edge of the nanoparticles. Additional information regarding the phase purity

Figure 13.1 Photograph showing modified combustion to form Y_2O_3-ZnO nanocomposite.

and the presence of any inorganic impurity was obtained using Fourier transform infrared spectrometer (FTIR) (Spectrum 2, PerkinElmer, Singapore) in the range 400–4000 cm^{-1} using ATR mode. Particulate properties of the combustion product are examined using high resolution transmission electron microscopy (HRTEM, Jeol/JEM 2100) at 200 kV. The samples for HRTEM were prepared by ultrasonically dispersing the powder in methanol and allowing a drop of this to dry on a carbon-coated copper grid. The thermal analysis of the sample was performed through thermogravimetric analysis technique and Differential thermal analysis (TGA/DTA) by (Perkin Elmer STA 6000) thermal analyzer in the range 15–900°C at a heating rate of 20°C min^{-1} in nitrogen atmosphere.

The high-quality ultrafine nanopowder is further compacted to a disc-shaped pellet in a steel die of 14 mm at 200 MPa pressure in a hydraulic pressing instrument. Sintering of the green body is carried out in a microwave sintering furnace (VBCC/MF/86, VB Ceramics Consultants, India). The microwave heating was realized using a pair of 2.45 GHz magnetrons with 1.1 kW. The green body is sintered to optimum density.

13.3 Results and Discussion

The powder XRD analysis technique is used to study the crystal structure of the as-synthesized sample. The XRD pattern for Y_2O_3-ZnO nanocomposite

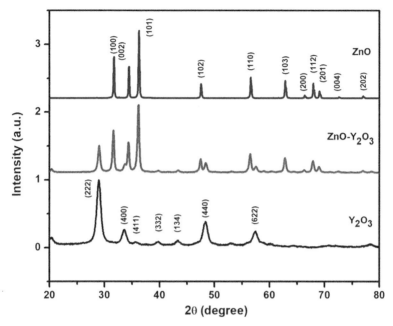

Figure 13.2a (a) XRD patterns of the as-prepared Y_2O_3-ZnO nanocomposite along with the ZnO and Y_2O_3 patterns.

at 50:50 mass % concentrations of yttria and zinc is shown in the Figure 13.2a. The XRD analysis reveals that the as-prepared powder is composed of cubic Y_2O_3 (ICDD:89-5591) and ZnO (ICDD:89-1397) phases, and the results confirm the formation of yttria and zinc oxide nanoparticles using the autoignited single-step modified combustion method. All the peaks in the pattern are indexed and match very well with the ICDD data. The sample shows a crystallite size of ∼16 nm for (222) peak of Y_2O_3 and ∼22 nm for (110) peak of ZnO. The crystallite size is found to be in the range of 12–26 nm, and the average crystallite size calculated is 19 nm. This result reveals that modified combustion technique is a simple and economic method for obtaining high-quality starting nanopowder in the exact and desired stoichiometry. Also, post annealing or calcinations is not applied for obtaining the voluminous final product.

The line broadening in the diffraction peak due to lattice strain caused by the non-uniform displacement of atoms with respect to lattice position can be explained by Hall–Williamson plot, Figure 13.2b shows the Hall–Williamson plot of the as-prepared nanoparticles. The reciprocal of the y intercept gives

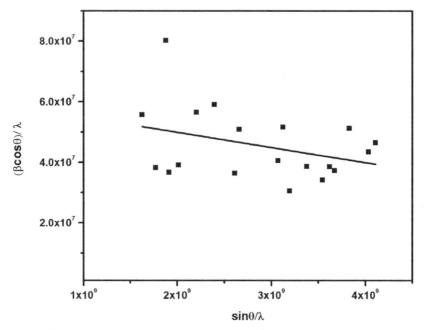

Figure 13.2b   (b) Hall–Williamson plot of the as-prepared Y_2O_3-ZnO nanocomposite.

the particle size of 16 nm. The lattice strain constant is proportional to the slope of the line and is estimated to be -5.01×10^{-3}. The negative slope of the line indicates the presence of compressive strain, and this may be due to lattice shrinkage effect [15, 16].

Figure 13.3a shows the HRTEM images of the as-synthesized nanocomposite sample with 50 mass % of yttria and 50 mass % of ZnO. The grains are distinctly visible in the HRTEM image. Figure 13.3b shows the crystallite size distribution of the as-prepared 50:50 mass % of Y_2O_3-ZnO sample. HRTEM image is shown in the inset. The crystallite size is found to be in the range of 12–26 nm, and the average crystallite size for the entire distribution is 19.8 ± 0.04 nm which corroborates with the XRD results. Figure 13.3c shows the selected area electron diffraction (SAED) pattern of the as-synthesized 50:50 mass % Y_2O_3-ZnO sample. The (222), (332) planes of yttria phase and the (100), (440) and (110) planes of ZnO phase are visible in the SAED pattern.

FTIR is an effective analytical tool for the identification of unknowns, sample screening, and profiling samples. Figure 13.4 shows IR transmittance of the selected 50:50 mass % Y_2O_3-ZnO sample in the range of 2.5–20 μm.

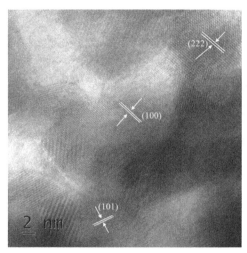

Figure 13.3a (a) HRTEM image of different crystallographic planes of 50:50 mass % Y_2O_3-ZnO sample.

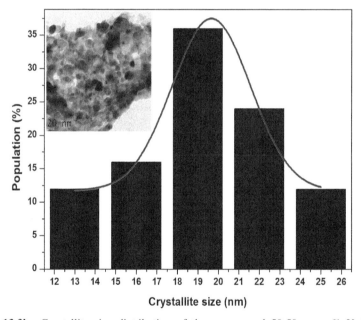

Figure 13.3b Crystallite size distribution of the as-prepared 50:50 mass % Y_2O_3-ZnO sample. HRTEM image is shown in the inset.

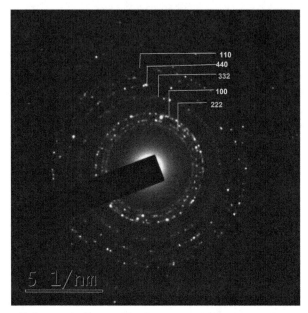

Figure 13.3c   (c) SAED pattern of the as-synthesized Y_2O_3-ZnO sample.

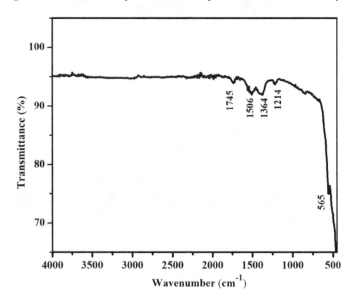

Figure 13.4   FTIR spectrum of the as-prepared sample with 50:50 mass % Y_2O_3-ZnO. Inset shows the expanded spectrum in the region 400–500 cm^{-1}.

The peak around 1745 and 1214 cm^{-1} represent the bending vibration of carbonate species. The peaks observed at 1506 and 1364 cm^{-1} are due to the asymmetrical and symmetrical stretching of the zinc carboxylate, respectively. The peak at 565 cm^{-1} corresponds to the metal oxide bonds of yttria. No other absorption peaks are observed in this range which shows organic matters and impurities are present in the starting powder [1, 17–20].

UV-visible spectroscopy explores the interaction between ultraviolet or visible electromagnetic radiation and matter. The spectrum obtained in the case is a plot of wave length of absorption peaks corresponding to the wavelength of radiation whose energy is equal to that required for an electronic transition. The absorption spectrum of the as-prepared samples of yttria nanoparticles was recorded using a spectrophotometer (UV-1700, Shimadzu, Singapore).

The UV-visible absorbance spectrum of the sample is recorded in the range of 200–800 nm, shown in Figure 13.5a. The sample shows maximum absorption in the UV of 200–410 nm and suddenly decreases at the visible region to almost 85%. The maximum absorption occurs at 410 nm. Thus, the material restricts the UV light to pass through it and hence finds application in the field of UV filters. This poor transmittance of hazardous UV light is

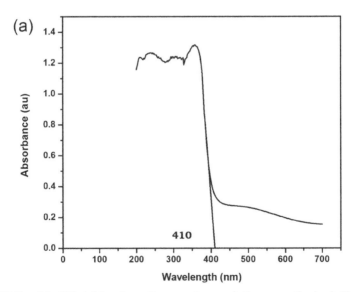

Figure 13.5a (a) UV-visible absorption spectrum of the as-synthesized Y$_2$O$_3$-ZnO nanocomposite.

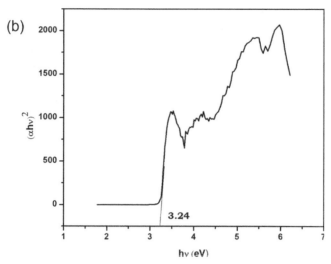

Figure 13.5b   (b) Tauc's plot of the as-synthesized ZnO-Y_2O_3 nanocomposite.

another foremost property which makes the sample an excellent material for infrared transparent material. The band gap of the Y_2O_3-ZnO nanocomposite was calculated using Tauc's plot extrapolating the curve drawn between (hν) and $(\alpha h\nu)^2$. The optical band gap energy was determined by extrapolating the wavelength of onset absorption in the UV region. Figure 13.5b shows Tauc's plot of the as-prepared Y_2O_3-ZnO sample. The band gap energy corresponding to 410 nm is found to be 3.24 eV. Thus, the sample is a direct wide band gap material that transmits wavelength in visible range, which confirms the suitability of sample to be used as an infrared transparent material [21, 22].

Thermal analysis measures physical or chemical changes in a material as a function of temperature. The thermal analysis of the sample with 50:50 mass % Y_2O_3-ZnO is taken. Figure 13.6 shows the TGA–DTA curves of the as-prepared 50:50 mass % Y_2O_3-ZnO in the temperature range from 35°C to 890°C at a heating rate of 20°C min^{-1} in nitrogen atmosphere. Even though no considerable weight loss was observed for the sample, a small change of 7% is due to the desorption or removal of moisture and solvents. This negligible weight loss discloses the purity and thermal stability of the sample. In the DTA curve, a small endothermic dip below 100°C occurs due to the evaporation of adsorbed water. In the DTA curve, the exothermic peak near 350°C is due to the presence of small amounts of organic matter in the

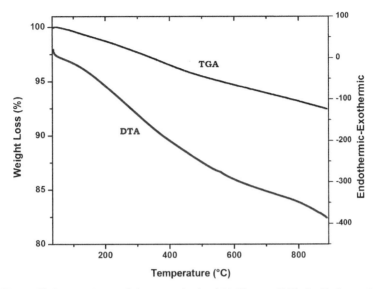

Figure 13.6 TGA/DTA of the as-synthesized 50:50 mass % Y_2O_3-ZnO sample.

sample. The absence of phase changes in the TGA–DTA curve reveals the thermal stability of the sample. The thermal stability of the material reveals its possibility to be used as an excellent infrared transparent ceramic material for which thermal stability is one of the foremost requirements [23, 24].

The green pellet with 55% of the theoretical density is further sintered to optimum density by a microwave sintering technique. The sintering technique is superior to the conventional sintering aids with reduced processing cycle, high production rate, and low cost and maintenance. The green body is densified to 96.2% of the theoretical density at 1410°C followed by a heating rate of 20°C/min at a soaking duration of 30 min which is much lower to pure yttria densified to ∼68% at the same temperature [1]. The reduction in sintering temperature is closely related to the ionic radius difference of matrix element Y^{3+} (0.9 Å) and additive element Zn^{2+} (0.6 Å), and it is influenced by the sintering method. In microwave method, the material is coupled with microwaves in such a way that the sintering is generated within the material initially and spreads to the entire volume. The variation of relative density with temperature is shown in Figure 13.7. The studies show that the quality of starting powder and the sintering method are the critical parameters for achieving transparency [25].

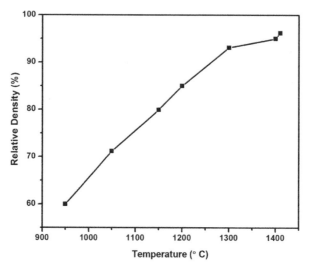

Figure 13.7   Variation of relative density of the sample with temperature.

13.4 Conclusions

Synthesis of high-quality Y_2O_3-ZnO nanocomposite powder by a modified combustion technique for the fabrication of infrared transparent ceramic material is presented in this chapter. The sample is synthesized at 50:50 mass percent concentration for tuning the physical properties and to yield superior quality IR transparent ceramics. The as-prepared sample is characterized using X-ray diffraction technique and confirms the formation of Y_2O_3-ZnO nanocomposite. Particulate properties of the combustion product are studied with the help of high-resolution transmission electron microscopy (HRTEM). The crystallite size and crystalline planes obtained from HRTEM are in good agreement with the XRD results. The phase purity of the sample is confirmed through FTIR. From UV-Visible spectrum it is clear that the material absorbs ultraviolet spectrum and transmits the visible light and hence used in the filters and sensors of UV radiations. The stability of Y_2O_3-ZnO nanocomposite sample is confirmed through thermal analysis. All the properties exhibited by the sample throughout the characterizations and high densification of the sample to 96.2% at a low sintering temperature attribute to the impressive quality of the starting powder. Thus, the single-step-modified combustion method is a simple, effective, and low-cost technique to synthesize high-quality ultrafine starting powder, and the microwave sintering technique is an additive to improve the performance of infrared transparent ceramic material.

Acknowledgement

The authors would like to acknowledge Mar Ivanios College, Thiruvananthapuram, and University of Kerala for providing the necessary facilities and resources for carrying out this research work.

References

[1] C. T. Mathew, Sam Solomon, Jacob Koshy, and Jijimon K. Thomas. Infrared transmittance of hybrid microwave sintered yttria. Ceramics International 2015, 41(8): 10070–10078.

[2] J. James, R. Jose, Asha M. John, and J. Koshy. A single step process for the synthesis of nanoparticles of ceramic oxide powders. U.S. Patent 6761866, 2004.

[3] Umar Al-Amani, S. Sreekantan, M. N. Ahmad Fauzi, A. R. Khairunisak, and K. Warapong. Soft combustion technique: Solution combustion synthesis and low-temperature combustion synthesis to prepare Bi4Ti3O12 powders and bulk ceramics. Science of Sintering 2012, 44: 211–221.

[4] R. V. Mangalaraja, P. Hedström, I. Kero, K. V. S. Ramam, Carlos P. Camurri, and M. Odén. Combustion synthesis of Y_2O_3 and $Yb–Y_2O_3$: Part I. Nanopowders and their Characterization 2008, 208: 415–422.

[5] E. L. Head and C. E. Holley. Modified combustion procedure for determining carbon and hydrogen in certain organometallic compounds. Anal. Chem. 1956, 7: 1172–1174.

[6] D. C. Harris. Durable 3–5 mm transmitting infrared window materials. Infrared Phys. Technol. 1998, 39: 185–201.

[7] R. K. Goyal, A. N. Tiwari, and Y. S. Negi. Microhardness of PEEK/ceramic micro- and nanocomposites: Correlation with Halpin–Tsai model.

[8] G. Zamfirova, V. Lorenzo, R. Benavente, and M. Jose, Perena. On the relationship between modulus of elasticity and microhardness. Journal of Applied Polymer Science 2003, 88: 1794–1798.

[9] Paola Palmero. Structural ceramic nanocomposites: A review of properties and powders' synthesis methods. Nanomaterials 2015, 5: 656–696.

[10] C. T. Mathew, S. Vidya, Jacob Koshy, Sam Solomon, and Jijimon K. Thomas. Enhanced infrared transmittance properties in ultrafine $MgAl_2O_4$ nanoparticles synthesized by a single step combustion method, followed by hybrid microwave sintering. Infrared Physics & Technology 2015, 72: 153–159.

[11] J. K. Thomas, H. Padma Kumar, S. Solomon, C. N. George, K. Joy, and J. Koshy. Nanoparticles of $SmBa_2HfO_{5.5}$ through a single step auto-igniting combustion technique and their characterization. Phys. Status Solid (a) 2007, 204(9): 3102–3107.

[12] J. I. Langford and A. J. C. Wilson. Scherrer after sixty years: A survey and some new results in the determination of crystallite size. J. Appl. Cryst. 1978, 11: 102–113.

[13] H. P. Klug and L. E. Alexander. X-ray diffraction procedures for polycrystalline andamorphous materials. Wiley 1974, p. 2.

[14] A. Patterson. The Scherrer formula for x-ray particle size determination. Phys. Rev. 1939, 56(10): 978–982.

[15] Anand Kumar Tripathi, Mohan Chandra Mathpal, Promod Kumar, Vivek Agrahari, Manish Kumar Singh, Sheo Kumar Mishra, M.M. Ahmad, and Arvind Agarwal. Photoluminescence and photoconductivity of Ni doped titania nanoparticles. Adv. Mater. Lett. 2015, 6: 201–208.

[16] A. Khorsand Zak, W. H. Abd. Majid, M. E. Abrishami, and Ramin Yousefi. X-ray analysis of ZnO nanoparticles by Williamson–Hall and size–strain plot methods. Solid State Sciences 2011, 13: 251–256.

[17] G. Xiong, J. G. Serrano, K. B. Ucer, and R. T. Williams. Photolumines-cence and FTIR study of ZnO nanoparticles: The impurity and defect perspective. Phys. Stat. Sol. 2006, 3: 3577–3581.

[18] Harish Kumar and Renu Rani. Structural and optical characterization of ZnO nanoparticles synthesized by microemulsion route. International Letters of Chemistry, Physics and Astronomy 2013, 14: 26–36.

[19] C. T. Mathew, Jijimon K. Thomas, Y. V. Swapna, Jacob Koshy, and Sam Solomon. Comprehensive analysis of the influence of resistive coupled microwaves sintering on the optical, thermal and hardness properties of infrared transparent yttria-magnesia composites. Ceramics International 2017, 43: 17048–17056.

[20] Morteza Hajizadeh-Oghaz, Shoja Razavi, Masoud Barekat, Mahdi Naderi, Saadat Malekzadeh, and Mohammad Rezazadeh. Synthesis and char-acterization of Y_2O_3 nanoparticles by sol–gel process for transparent ceramics applications. Journal of Sol-Gel Science and Technology 2016, 78: 682–691.

[21] Tong-ming Su, Zu-zeng Qin, Hong-bing Ji, and Yue-xiu Jiang. Prepa-ration, characterization, and activity of Y_2O_3-ZnO complex oxides for the photodegradation of 2,4-dinitrophenol. International Journal of Photoenergy 2014. http://dx.doi.org/10.1155/2014/794057.

[22] B. N. Lakshminarasappa, J. R. Jayaramaiah, and B. M. Nagabhushana. Thermoluminescence of combustion synthesizes yttrium oxide, Powder Technol. 2012, 217: 7–10.

[23] Yendrapati Taraka Prabhu, Kalagadda Venkateswara Rao, Vemula Sesha Sai Kumar, and Bandla Siva Kumari. Synthesis of ZnO nanoparticles by a novel surfactant assisted amine combustion method. Advances in Nanoparticles 2013, 2: 45–50.

[24] S. Ghorbani, M. R. Loghman-Estarki, R. Shoja Razavi and A. Alhaji. A new method for the fabrication of MgO-Y_2O_3 composite nanopowder at low temperature based on bioorganic material. Ceramics International 2018, 44: 2814–2821.

[25] Mohammad Reza Arefi and Saeed Rezaei-Zarchi. Synthesis of zinc oxide nanoparticles and their effect on the compressive strength and setting time of self-compacted concrete paste as cementitious composites. International Journal of Molecular Sciences 2012, 13.

MODULE 4

14

Atomically Precise Fluorescent Metal Nanoclusters as Sensory Probes for Metal Ions

Shilpa Bothra and Suban K. Sahoo*

Department of Applied Chemistry, SV National Institute of Technology (SVNIT), Surat-395007, Gujarat, India
E-mail: suban_sahoo@rediffmail.com; sks@chem.svnit.ac.in
*Corresponding Author

Fluorescent metal nanoclusters (NCs) gained a bourgeoning interest in the field of nanomaterials due to the special properties of ultra-small size, low toxicity, excellent photostability and bio-compatibility. With the size approaching to the Fermi wavelength of electrons, metal NCs possess molecule-like properties and excellent fluorescence emission. As the photophysical properties of such NCs materials depend strongly on size, focus is laid on development of synthetic routes that allows precise tuning of the cluster cores with high monodispersity and purity. Various small molecules, proteins, polyelectrolytes etc. are used for the direct one-pot synthesis of noble metal fluorescent nanoclusters and applied for the sensing of metal ions including other analytes. The novel optical properties render the fluorescent noble-metal nanoclusters as ideal fluorophores for multicolor and multiplexing applications in biomedical engineering and molecular biotechnology. This chapter provides an overview of the properties, synthesis and their applications for the detection of metal ions.

14.1 Introduction

The colorimetric and fluorimetric techniques employed for the detection of chemically and biologically important analytes like metal ions, anions and neutral molecules have keenly attracted scientists including chemists, biologists, nanotechnologist, clinical biochemists and environmentalists in the last few decades due to their high selectivity, sensitivity and the real-time monitoring with fast response time [1]. Metal ions are involved in cellular and subcellular functions [2] and can effectively control an enzyme-catalyzed reaction [3]. Thus, the consumption of essential (Cu^{2+}, Zn^{2+}, Mg^{2+}, Fe^{3+} etc.) and non-essential (Cd^{2+}, Hg^{2+}, Al^{3+} etc.) metal ions in appropriate proportion into the body becomes vital as its disproportion at relatively high concentrations can cause threat to human health through cellular toxicity, liver damage and kidney damage, and neurodegenerative diseases and so on [4]. It is therefore essential to monitor these influential metal ions in the environment, drinking water, food, and biological fluids.

Conventional methods for heavy metal measurement include atomic absorption spectroscopy, inductively coupled plasma/mass spectrometry, inductively coupled plasma/atomic emission spectrometry, ultraviolet-visible spectroscopy, etc. Although these techniques are highly sensitive and selective, they require tedious sample preparation and pre-concentration procedures, expensive instruments, and professional personnel [5]. Moreover, they cannot be used as portable devices for on-site detection. Therefore, it is of great concern to develop a simple, feasible and reliable method for the selective qualitative and quantitative detection of these biologically relevant ions. Thus, nanomaterials provide novel systems for the pursuit of new recognition and transduction processes, as well as new opportunities for improving the performance of sensors in terms of sensitivity, limit of detection, selectivity, and reproducibility, and increasing the signal-to-noise ratio through miniaturization with assistance of lab-on-chip (LOC) technology. This successful rapidly expanding research in nanotechnology is due to the ability of researchers to manipulate the desired properties of nanomaterials by controlling their size, shape, and composition.

Recently, fluorescent nanomaterials have attracted a great deal of attention in the past few decades over organic dyes owing to anti-photobleaching, high specific surface area, easy modification, and color tenability, all of which provide promising platforms for a wide variety of applications, including chemical sensing, bioimaging and catalysis, and in electronic devices [6]. Numerous fluorescent nanomaterials including semiconductor quantum dots, metal nanoclusters (NCs), polymer dots, up-conversion nanoparticles (NPs), dye-doped NPs, and noble metal based nanoparticles and nanoclusters have keenly attracted interest of researchers working in the field of colorimetric and fluorescent sensors development owing to their unique optical properties [7]. Tremendous emphasis is laid to nanomaterials over the bulk materials due to variation in the redox properties, band gap, enhancement in toughness and strength, anomalous melting points and unusual crystal structures (in metals) which results from small sizes, including quantum size effect on photochemistry, non-linear optical properties of semiconductor or the emergence of metallic properties with the size of the particles [8]. Also, small size and high surface area of nanomaterials allows further modification with hydrophobic, hydrophilic, cationic, anionic or any neutral moieties to the surrounding environment, enabling many applications in biological sciences.

14.2 Fluorescent Metal Nanoclusters

Among the varied types of nanomaterials, metal NCs have especially been recognized as the rising stars in several technological areas, spanning from solid state lighting, solar cells and sensors to photo-catalysis and biomedical applications owing of their size- and shape-tunable electronic properties, ultra-large surface-to-volume ratios, low toxicity and the flexibility of their physical properties *via* surface functionalization [9]. Naming of these materials is still a point of debate and a fully acceptable terminology has not appeared so far. As a result, authors use a number of names, which include nanomolecules, nanoclusters, NPs, faradaurates, monolayer-protected clusters (MPCs), artificial atoms, super atoms, QCs, etc. These nanoclusters are unique, in that they can be represented with definite formulas (as opposed to an average size with size distribution), resembling well-defined organic molecules or organometallic compounds; thus, nanoclusters are essentially inorganic–organic hybrid molecular compounds. Metal nanoclusters are defined as isolated particles consisting of several to tens of metal atoms, typically have a size smaller than 2 nm, and have been attracting attention for their unique role in bridging the "missing link" between atomic and nanoparticle behaviour as shown in Figure 14.1a. Metal nanoparticles (>2 nm) exhibit quasi-continuous energy levels and exhibits intense colors due to the collective oscillation of conduction electrons upon interaction with light, in short named as surface plasmon resonance (SPR). However, when their dimension is further reduced to the size approaching the Fermi wavelength of electrons, the band structure shows discrete energy levels. For instance, gold nanoparticles show a size-dependent plasmon absorption band, when their conduction electrons in both the ground and excited states are confined to dimensions smaller than the electron mean free path (ca. 20 nm),

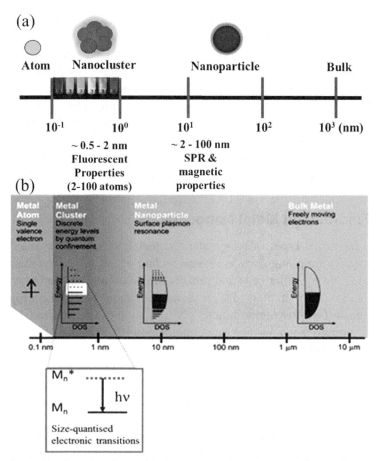

Figure 14.1   (a) Hierarchy of materials from atom to bulk. (b) The effect of size on metals. Bulk metal and metal nanoparticles possess a continuous band of energy levels. While the limited number of atoms in metal nanoclusters results in discrete energy levels, allowing interaction with light by electronic transitions between energy levels. Metal nanoclusters bridge the gap between single atoms and nanoparticles [11].

but plasmon absorption disappears completely for nanoparticles less than 2 nm where Mie's theory no longer can be applied. Thus, metal nanoclusters shows confinement to a second critical regime having sizes comparable to the Fermi wavelength of the electron (ca. 0.7 nm), which displays molecule-like properties of discrete electronic states and size-dependent fluorescence i.e., a scale function of the number of atoms within the cluster from the energy differences between the highest occupied molecular orbital (HOMO) and the lowest unoccupied molecular orbital (LUMO) (Figure 14.1b). These energy transitions can be rationalized according to the jellium model ($E_{fermi}/N^{1/3}$) [10]. The fluorescence emission of metal NCs are highly sensitive to their chemical environment, including the metal cluster, solvent, and surface protecting ligands.

The two possible mechanisms that could be used to explain the observed emission from these small metal clusters, intraband transition of the lowest unoccupied molecular orbital to the highest occupied molecular orbital and inter-band transition between the 6 sp conduction band and the filled 5d band. The fluorescence originating from the core tends to follow the jellium quantum mechanics model which is based on the Drude free electron models where "magic" clusters (i.e., clusters of remarkable structural, electronic and thermodynamic stability) are a result of the complete filling of the different valence shells. Metal nanocluster of certain number of atoms possessing extraordinary stability originating from either atomic or electronic shell closing are known as magic clusters [12]. Most of the metal clusters shows following series of magic numbers: 2, 8, 18, 20, 34, 40, 58 etc. [13]. Although the crystal structure remains intact at nano scale for most of the metals, understanding the crystal structure of metal nanoparticles becomes very essential to explain the chemistry on their surfaces. Usually the metals possess cubic lattices. Typical elements such as Cu, Ag, Au, Ni, Pd, Pt, and Al crystallizes in the face centred cubic (FCC) structure whereas Fe, Cr, V, Nb, Ta, W and Mo shows the body centred cubic (BCC) crystal structure.

Also, several groups researching on the emission mechanism suggests that there may be two major fluorescent sources for metal NCs. One source is the metal core with its intrinsic quantization effects. The emission in the visible region is fast and very short-lived (hundreds of fs). The other source is the particle surface governed by the interaction between metal core and surface ligands, and the emission is long-lived (up to μs) and near-infrared fluorescence [14]. Although, the exact mechanism of metal NCs remains unknown and unexplored, it has been suggested that their luminescence is highly dependent on the size of the metal cores and surface ligands [15].

Also, the strong quantum size effects in nanoclusters are manifested in their physicochemical properties such as catalysis [16], chirality [17], magnetism [18], antimicrobial activity [19] and non-linear optical properties [20]. Because of a greater proportion of high energy orbitals, metal NCs have been considered as active catalysts in recent years.

In the past few decades, the numerous metal NCs with excellent fluorescence emission ability including gold (Au) [6, 21], silver (Ag) [21, 22],

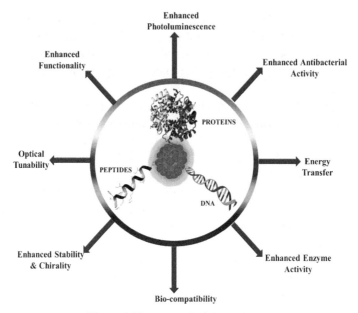

Figure 14.2   Properties of metal NCs.

copper (Cu) [21, 23], platinum (Pt) [21, 24], bismuth (Bi) [25], molybdenum (Mo) [26] or mixed metals [27] has been reported. Precisely, noble metal nanoclusters (e.g., Au, Ag) have been attracting attention as a new type of luminescent probes for the development of optical sensors due to strong luminescence with ultra-small size, good optical and colloidal stability, and well-defined composition and structure (Figure 14.2). Gold (Au) and silver (Ag) popularity in nanomaterials and self-assembled monolayer research is largely based on its relative inertness (e.g. resistance to oxidation) and strong binding affinity for thiolates [28].

14.3 Synthesis of Fluorescent Metal Nanoclusters

Creation of clusters with varying core sizes and ligands is fascinating as it opens up immense opportunities to study the emergence of novel physical and chemical properties and provides an insight into their structure–property relationships. Just similar to the synthesis of larger metal NPs, the synthesis of metal NCs can generally be classified into two main groups: top-down and bottom-up approach (Figure 14.3). The top-down synthetic routes include mainly size focusing methods on the basis of ligand-mediated etching from

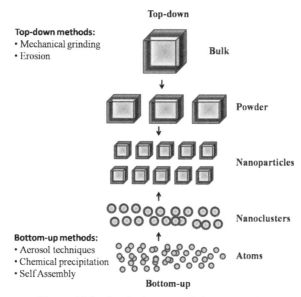

Figure 14.3 Synthesis approach for metal NCs.

larger NPs [29] or clusters [30] can be obtained as a result of the strong interaction between metal atoms and protecting ligands. This core etching of the metallic NPs is generally done using etching molecules such as thiols, alkylated thiols, amines, phosphines, and polymers [31]. Solvent etching or phase-transfer is another suitable top-down method to prepare fluorescent NCs *via* electrostatic interaction. This synthetic routes produces fairly monodisperse clusters of definite nuclearity upon careful control of precursor metal ion and ligand concentration. Also, this size focusing methods may provide selective synthetic routes for a number of atomically monodispersed ultra-small metal NCs owing to the peculiar stability of certain-sized metal NCs.

The bottom-up synthetic routes consist mainly of Brust-Schiffrin methods, template-based synthesis methods, microwave-assisted synthesis, electrochemical, sonochemical and photo-reduction synthesis (Figure 14.4) [32]. In a typical bottom-up synthetic process, the metal precursors are reduced to zero-valent metal atoms using a certain reducing agent, and the metal clusters (Au, Ag, Pt, Cu) with high stability and quasi-monodisperse size distribution are generated. The properties of metal NCs can easily be controlled through the adjustment of experimental parameters, such as the surface ligands, the ratio of the metal to the ligand, reducing agent, reaction time,

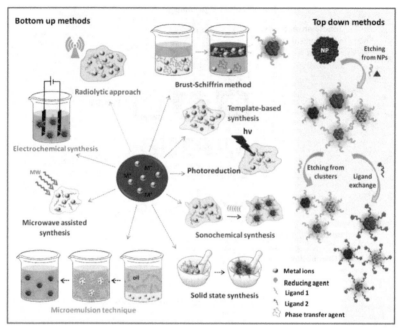

Figure 14.4 Various routes for the synthesis of atomically precise, sub-nanometer-sized, noble metal quantum clusters [41].

and medium pH. Owing to their ultra-small size and high surface energy, bare metal NCs tend to aggregate to form larger NPs without any fluorescent signal. Metal nanoclusters have also gained great improvement in synthesis strategies which can be divided into two categories: the template-based synthesis and monolayer-protected nanoclusters (MPCs) synthesis. Template-assisted synthesis uses biomolecules (e.g., proteins and DNA), dendrimers (e.g., poly(amidoamine) (PAMAM)), and polymers (e.g., poly(methacrylic acid) (PMAA)) are as a template to guide the formation of fluorescent Au and/or Ag NCs and to prevent aggregation by reducing the surface energy. However, the as-synthesized metal NCs are embedded in the template molecules, leading to a relatively large hydrodynamic diameter (>3 nm), which may affect their usability as a fluorescent label for some specific small molecules or few-nanometer-sized biomolecules. The second approach relies on the use of specific capping agents, e.g., thiol ligands, which interact strongly with the noble metal surface to form ultra-fine-sized metal NCs protected by a monolayer of thiol ligands. Thiol-protected metal NCs have attractive features such as excellent stability, ultrasmall hydrodynamic diameters, and modifiable surface properties. Numerous fluorescent Au and Ag NCs were synthesized using several mercapto linkers such as dodecanethiol, mercaptoundecanoic acid, thiolate-α-cyclodextrin, or D-penicillamine [33]. Recent studies have shown that highly fluorescent Au and Ag NCs can be synthesized via the etching of large metal nanoparticles (>3 nm) by excess ligands. Thus, emphasis is laid on developing a facile and scalable method to fabricate thiol-protected fluorescent NCs of different metals, including Ag, Au, Pt, and Cu, to expand the repertoire of this potential new class of bio-imaging probes, and to investigate composition/structure property relationships. Moreover, fluorescent Ag NCs with different functional groups on their surface (e.g., carboxyl, hydroxyl, and amine) can be prepared by the phase transfer method using simple custom-designed peptide ligands with the desired functionalities.

Various proteins, such as bovine serum albumin (BSA) [34], human serum albumin (HSA) [35], insulin [36], horseradish peroxidase [37], pepsin [38], lactotransferrin [39], as well as lysozyme [40] have been employed as templates for the preparation of the fluorescent metal NCs. Protein-directed synthesis is particularly very attractive because proteins serve as environmentally-benign reducing and stabilizing molecules, require only mild reaction conditions, and offer great water solubility and natural bio-compatibility. Furthermore, the 3D complexed structures of proteins can withstand a wide range of pH and can be easily conjugated with other. Thiolated, carboxylic and amine terminated ligands are routinely used as protecting and stabilising agent for synthesis and stabilization of highly fluorescent NCs due to their unique and strong interaction with noble metals, which could lead to an extraordinary stability and distinct properties of pho-toluminescence. This labeling of biomolecule of fluorescent metal NCs is the key basis for their further application to specific biosensing and bio-imaging which is commonly based on passive adsorption, multivalent chelation and covalent-bond formation.

Bio-conjugation of colloidal nanoparticles is the 'natural' extension of the described concepts of ligand exchange and chemical functionalization to biomolecules. It simply involves the bond of biomolecules to nanoclusters by chemical or biological means, which render them ideal for biological and environmental applications; it includes the conjugation of biologically active molecules to nanomaterials. The conjugation of different function-alized groups to nanomaterials is necessary for their stability, functionality and biocompatibility by using reactive functional groups of primary amines, carboxylic acids, alcohols, or thiols and develops their application fields, and provides them with novel and improved properties. Few of the examples includes small molecules like lipids, vitamins, peptides, sugars and larger ones such as natural polymers including proteins, enzymes, DNA and RNA. Nanomaterial–biomolecule conjugation brings unique properties and functionality of both materials, e.g. fluorescence or magnetic moment of the inorganic particles and e.g. the ability of biomolecules for highly specific binding by molecular recognition. Mostly fluorescent gold nanoclusters are protected by the carboxylic acids, which can conjugate with amino-molecules to form stable amide bond catalyzed with a carbodiimide or sulfo N-hydroxysuccinimide ester. Commonly found carboxylic groups can be reacted with primary amines by means of a condensation reaction to yield amide bonds. For this, a water-soluble carbodiimide (e.g. EDC) is commonly used. The fluorescence of NCs with the advantage of bioconjugation with small molecules, drugs, proteins etc. led to the wide applications in bioimaging, detection, delivery agents etc.

14.4 Detection of Metal Ions with Fluorescent Metal Nanoclusters

Unique properties of clusters such as large Stokes shift, biocompatibility, low toxicity, ease of conjugation, as well as size and ligand dependent fluorescence properties enable them to be developed into useful materials in energy, environment, and biology. A strong luminescence or high QY of the NCs is crucial for realizing a good sensitivity of the fluorescent sensors. The luminescence properties (e.g., emission intensity and wavelength) of the NCs are highly sensitive to the local environment, the size and structure of the NCs that provide an excellent response for signalling their interaction with analytes. "Turn-off" and "Turn-on" luminescence detection are two common schemes in the NCs-based optical sensors based on aggregation-induced emission. In a typical "turn-off" scheme, the luminescence of the NCs is quenched by the analyte due to the specific interaction between the analyte and NCs (via the metal core or ligand shell). In a typical "turn-on" scheme, the luminescence of the NCs is initially annulled by an inhibitory agent [e.g., concanvalin A], the addition of the analyte can selectively remove the inhibitor from the NCs, restoring the luminescence of the NCs. Both metal core and ligand shell of the NCs can serve as the recognition component for an analyte. Three primary interactions between the metal (Au/Ag) core and analytes have been reported: metallophilic interactions, analyte deposition on the metal core surface, and analyte-induced metal core decomposition. With these interactions, highly luminescent Au and Ag NCs have been used to construct fluorescent sensors for a variety of chemical and biological analytes, including cations, anions, small organic molecules, and biomolecules.

Heavy metal contamination in drinking water is a major problem faced by the global community. Heavy metal ions such as Hg^{2+}, Cd^{2+}, Pb^{2+}, and Cu^{2+} bind to various cellular components, such as proteins, enzymes, and nucleic acids, leading to alteration of their biological functions, which may cause serious diseases and death. Because of their high toxicity, the maximum allowed limits of Hg^{2+}, Cd^{2+}, Pb^{2+}, and Cu^{2+} in drinking water set by the environmental protection agency (EPA) of the United States are 0.002 (10), 0.005 (45), 0.015 (72), and 1.3 (21) ppm (nM), respectively. It is difficult to detect these ions in complicated biological and environmental samples. Thus, highly sensitive and selective sensing systems are required for their determination [42]. Several groups have reported that metal NCs can be applied as promising optical probes for the sensing of metal ions in water. Many fluorescent NCs as sensitive sensing probes have been used to detect various heavy metal ions, such as Zn^{2+}, Ag^+, As^{3+}, Cd^{2+}, Cr^{3+}, Cr^{6+}, Cu^{2+}, Fe^{3+}, Hg^{2+} etc. based on fluorescence quenching or enhancement.

As a highly toxic contaminant, Hg^{2+} can accumulate in the human body through the food chain and damage the brain, heart, kidney, stomach, and intestines, even at a very low concentration. Fluorescent gold nanoclusters (AuNC@MUA) were firstly used to sense mercury(II) based on fluorescence quenching through Hg^{2+}-induced aggregations of AuNC@MUA [43]. Lin et al. and Xie et al. have reported the sensitive detection of Hg^{2+} and CH_3Hg^+ based on the quenching of AuNCs by taking advantage of the $5d^{10}-5d^{10}$ interaction between Hg^{2+} and Au^+ [44]. Many sensitive and selective red fluorescent templated AuNCs systems have been developed for the detection of Hg^{2+} [40a, 45]. Denatured BSA stabilized Ag NCs (dBSA-Ag NCs) can also be used in the highly sensitive and selective detection of Hg^{2+} based on the $5d^{10}(Hg^{2+})$-$4d^{10}(Ag^+)$ metallophilic interaction with LOD of 10 nM, which achieves the required sensitivity for Hg^{2+} detection in drinking water permitted by the EPA [46]. This approach has the advantages of high stability, self-standing ability, naked-eye detection, selectivity, reproducibility, and easy handling.

Cu^{2+} is a significant environmental pollutant and an essential trace element in humans. When its amount reaches about 0.01 mg/L, Cu^{2+} has a significant inhibitory effect on the self-clarification ability of water, and is very harmful to aquatic organisms. GSH-Au NCs showed selective detection for Cu^{2+} with low limit of detection as 86 nM, based on analyte induced fluorescence quenching [47]. Numerous other NCs system such as mixture solution of BSA-Au NCs and lysine-stabilized Au NCs [48], Poly (methacrylic acid) (PMAA)-templated Ag NCs (PMAA-Ag NCs) [49], CdTe/Silica/Au NCs hybrid spheres [50], DNA Ag NCs [51], have been developed with simple and sensitive method for the determination of Cu^{2+} with low detection limit via 'turn OFF' or 'turn ON' mechanism. Further, copper has also been detected by rapid sonochemical synthesis of highly luminescent BSA-AuNCs [52].

Dihydroxyphenylalanine-capped Au NCs has also been applied to the detection of Fe^{3+} based on aggregation resulting due to o-Quinone-containing ligands complexation with Fe^{3+} leading to fluorescence quenching. Halawa et. al have recently reported the detection of Fe^{3+} by utilizing reducing-cum-stabilizing inositol fluorescent gold nanoclusters with limit of detection of 0.54 μM which is 10 times lower than the limit value ($\approx$5.5 μM) allowed by the U.S. Environmental Protection Agency in drinkable water, indicating the sensitivity of the fluorescent probe of Fe^{3+} sensing [53]. Recently, Huang et al. developed a simple method for the synthesis of albumin chicken egg capped porous copper nanoclusters (p-Cu NCs) for the first time which was successfully applied for the fluorescence sensing of Fe^{3+} ions [54].

New type of fluorescent chemodosimeter copper nanoclusters functionalised with cysteamine (Cys-CuNCs) has been demonstrated by Boonmee et al. for the turn-On detection of Al^{3+} ions with a low detection limit of 26.7 nM and was applied to determine Al^{3+} in drinking water samples with satisfactory results [55]. Hu et al. shows interaction of Al^{3+} ions leading to aggregation-induced fluorescence enhancement of the orange fluorescent CuNCs prepared by direct reduction of Cu(II) ions by dithiothreitol serving as protecting agent [56].

Recently, Lin et al. has reported photoluminescence light-up method for detecting Zn^{2+} based on the aggregation induced emission enhancement of GSH-capped Cu NCs. The PL of Cu NCs was enhanced after the addition of Zn^{2+} due to the formation of aggregates which restrained the intra-NCs or inter-NCs movements, inhibiting the radiation less relaxations and resulting in the increase of radiative decays [57].

Numerous other Ag/Au/Cu based metal nanoclusters templated with biomolecules such as proteins, peptides, enzymes etc. have been reported for the onsite detection of different metal ions with high selectivity, sensitivity and low limit of detection. In our approaches, we have developed noble metal (Au/Ag) nanoclusters based nanosensors by conjugating with vitamin B_6 cofactors like pyridoxal and pyridoxal 5′-phosphate. The vitamin B_6 cofactor plays crucial roles in enzymatically catalysed transaminations to form α-keto acid and pyridoxamine 5-phosphate as well as in many other biosynthetic processes [58]. Pyridoxal containing enzymes are central to numerous metabolic pathways such as decarboxylations of amino acids [59], racemization of amino acids [60] and aldol type addition of the pyridoxal stabilized glycine carbanion to formaldehyde or acetaldehyde [61]. Another vitamin B_6 cofactor PLP, one of the catalytically active forms of vitamin B_6, influence brain function by participating at stages in metabolism of proteins, lipids, carbohydrates, other coenzymes and hormones. Additionally, PLP helps to balance the sodium and potassium levels by regulating the electrical functioning of nerves, heart and musculoskeletal system. Also, in the catalytic process of PLP-dependent enzymes, the substrate amino acid forms a Schiff base with PLP and the electrophilicity of the PLP pyridine ring plays important roles in the subsequent catalytic steps [62]. This ease of vitamin B_6 cofactors to form various Schiff base derivatives to detect metal ions has led our research group to work on design and development of vitamin B_6 cofactor (pyridoxal/Pyridoxal 5 Phosphate) conjugated fluorescent nanoclusters for highly specific binding by molecular recognition [63].

Thus, considering the merit of ease of conjugation of vitamin B_6 with the amine or carboxylic group of fluorescent Au/Ag nanoclusters, we recently developed a fluorimetric method for distinct recognition of Hg^{2+} ions using pyridoxal (Py) conjugated BSA-AuNCs with the detection limit down to 31.9 nM as shown in Figure 14.5. The fluorescent system was developed by the simple Schiff base interaction between the amino groups of BSA-AuNCs and the aldehyde group of pyridoxal. Upon interaction with Hg^{2+}, the fluorescence of AuNCs was quenched and the naked-eye detectable red fluorescent color (under UV light) change to blue. The quenching could be attributed due to the chelation of heavy-atom Hg^{2+} with the imine-N and pyridoxal-O due to the photo-induced electron transfer followed by the formation of metallophilic bonding between Hg^{2+} ($4f^{14}5d^{10}$) and Au^+ ($4f^{14}5d^{10}$) due to the high specificity of Hg^{2+}–Au^+ interactions [64]. The developed pyridoxal conjugated BSA-AuNCs system was also applied on chemically-modified cellulosic paper strips for detecting Hg^{2+} ion up to 1 nM.

Similar to the Py_BSA-AuNCs approach, red fluorescent Lyso-AuNCs were developed and conjugated with the vitamin B_6 cofactor pyridoxal-5′-phosphate (PLP) via formation of a Schiff base between the –CHO group of

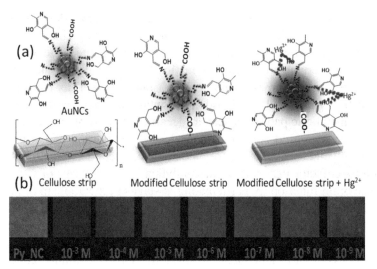

Figure 14.5 (a) Schematic representation for preparation of the paper-based device and (b) fluorescence color changes visualized on test paper strips of Py_BSA-AuNCs upon interaction with different concentrations of Hg^{2+} (1 mM to 1 nM) observed under UV light at 365 nm [64].

PLP with the free $-NH_2$ present in the lysozyme. The yellow fluorescence of PLP_Lyso-AuNCs turned to bluish-green fluorescence upon interaction with Zn^{2+} and this switching behavior causing enhancement for Zn^{2+} may be due to the coordinative binding of Zn^{2+} with the Schiff base resulted from the interaction of Lyso-AuNCs with the aldehyde group of PLP. With this system, the Zn^{2+} can be detected down to 39.2 nM and were successfully applied to monitor the intracellular Zn^{2+} in live HeLa cells as shown in Figure 14.6 [65].

Subsequently, the blue fluorescent PEI passivated AgNCs (PEI-AgNCs) were synthesised by the silver mirror reaction using formaldehyde as a reducing agent. The vitamin B_6 cofactor was conjugated over the surface of the fluorescent PEI-AgNCs and applied for the fluorescent turn-on sensing of Zn^{2+} and Cd^{2+} ions in solution and by using the chemically-modified cellulosic strips (Figure 14.7). The PLP conjugation was achieved due to the formation of Schiff base upon interaction with the aldehyde group of PLP and the free amines present in the PEI-AgNCs. This PLP conjugated nanoprobe showed a complexation-induce turn-on fluorescent response in the presence of Zn^{2+} and Cd^{2+} with the detection limit down to 50.5×10^{-8} M for Zn^{2+} and 59.0×10^{-8} M for Cd^{2+}, respectively. This reversible fluorescent nano-assembly with remarkable sensitivity have been successfully employed for

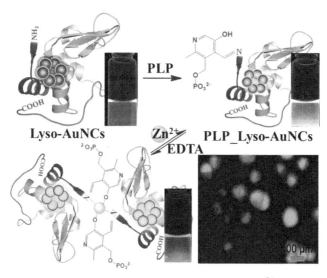

Figure 14.6 Schematic representation showing interaction of Zn^{2+} ions with conjugated Lyso-AuNCs and its intracellular detection (Adapted from Ref. [65]).

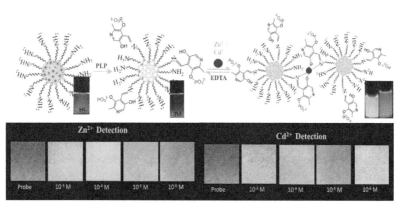

Figure 14.7 Schematic representation showing the mechanism for interaction of Zn^{2+} and Cd^{2+} with developed conjugated nanoprobes PLP-PEI-AgNCs and photographic images of the DVS modified cellulosic strips upon interaction with different concentrations of Zn^{2+} (10^{-3} M to 10^{-6} M) and Cd^{2+} (10^{-3} M to 10^{-6} M) observed under UV light at 365 nm [66].

the real samples analyses in the quantitative detection of Zn^{2+} and Cd^{2+} in various environmental and biological fluids with satisfactory recovery. The nanoprobe was covalently anchored over the cellulose strips by using the DVS linker and applied for the fluorescent detection of Zn^{2+} and Cd^{2+} [66].

14.5 Conclusions

The new electronic and optical properties of nanoclusters, as well as their unique atom-packing structures will have far-reaching impacts on the conceptual advances in materials design for nanocatalysis, sensing, storage and utilization of light and chemical energy. Given the potential applications of fluorescent metal nanoclusters in many fields, we believe the work described herein may also serve as a foundation for the further design of novel conjugated materials for possible applications in the field of immunoassays, fluorescence imaging, and unfolding biological processes and so on. Considering the outcomes on the excellent selectivity and low detection limit of the Vitamin B_6 conjugated nanoclusters in aqueous medium, this approach could potentially accelerate the design of many more fluorescent nanosensors using the bioactive vitamin B_6 cofactors and nanoclusters based on Au, Ag, Cu, etc. Therefore, a wide scope is open to further investigations on the applications of conjugated nanoclusters in the field of sensing. Precisely, these NCs functionalised with vitamin B_6 cofactors provide high specificity, sensitivity, low limit of detection, low cost of preparation and ease of synthesis. Fabrication of cellulosic strips is key advantage of these nanosensors as these could produce miniaturized equipment's utilizing optical characteristics. With the low or non-toxic components (noble Au/Ag cores and proteins/polyamines-like coating layers), small dimension for the distribution to subcellular domains, versatile surface chemistry for specific targeting (biomarker oriented), wide range for excitation wavelength, near-infrared emission, and longer life-time than auto fluorescence, we believe these conjugated luminescent NCs have great potentials in fluorescence cell imaging applications *in-vitro* and *in-vivo* applications. In addition, keeping the multiple reactive sites in the vitamin B_6 cofactors, many more conjugated systems can be developed utilizing the surface of different fluorescent metal nanoclusters for the detection of bioactive ions and neutral molecules. Therefore, the present review open a new direction for the utilization of the vitamers of vitamin B_6 in the field of nanosensing.

Acknowledgement

We would like to thank the Director, SVNIT for providing necessary research facilities and infrastructure. This work was made possible by a grant from the DST, New Delhi (SR/S1/IC- 54/2012).

References

[1] (a) Lee, J. H.; Lim, C. S.; Tian, Y. S.; Han, J. H. and Cho, B. R., A Two-Photon Fluorescent Probe for Thiols in Live Cells and Tissues. *J Am Chem Soc* **2010**, *132*(4), 1216–1217; (b) Wei, T.-B.; Zhang, P.; Shi, B.-B.; Chen, P.; Lin, Q.; Liu, J. and Zhang, Y.-M., A highly selective chemosensor for colorimetric detection of Fe3+ and fluorescence turn-on response of Zn2+. *Dyes and Pigments* **2013**, *97*(2), 297–302; (c) Chhatwal, M.; Kumar, A.; Singh, V.; Gupta, R. D. and Awasthi, S. K., Addressing of multiple-metal ions on a single platform. *Coord Chem Rev* **2015**, *292*, 30–55; (d) Nandre, J.; Patil, S.; Patil, V.; Yu, F.; Chen, L.; Sahoo, S.; Prior, T.; Redshaw, C.; Mahulikar, P. and Patil, U., A novel fluorescent "turn-on" chemosensor for nanomolar detection of Fe(III) from aqueous solution and its application in living cells imaging. *Biosens Bioelectron* **2014**, *61*, 612–617.

[2] Anastassopoulou, J. and Theophanides, T., The Role of Metal Ions in Biological Systems and Medicine. In *Bioinorganic Chemistry: An Inorganic Perspective of Life*, Kessissoglou, D. P., Ed. Springer Netherlands: Dordrecht, **1995**, 209–218.

[3] JJRFD, S. and Williams, R., The biological chemistry of the elements. Clarendon Press, Oxford: 1991.

[4] (a) Sahoo, S. K.; Sharma, D.; Bera, R. K.; Crisponi, G. and Callan, J. F., Iron(iii) selective molecular and supramolecular fluorescent probes. *Chemical Society Reviews* **2012**, *41*(21), 7195–7227; (b) Hamilton, G. R. C.; Sahoo, S. K.; Kamila, S.; Singh, N.; Kaur, N.; Hyland, B. W. and Callan, J. F., Optical probes for the detection of protons, and alkali and alkaline earth metal cations. *Chemical Society Reviews* **2015**, *44*(13), 4415–4432; (c) Baral, M.; Sahoo, S. K. and Kanungo, B. K., Tripodal amine catechol ligands: A fascinating class of chelators for aluminium(III). *Journal of Inorganic Biochemistry* **2008**, *102*(8), 1581–1588.

[5] Li, M.; Gou, H.; Al-Ogaidi, I. and Wu, N., Nanostructured sensors for detection of heavy metals: a review. ACS Publications: 2013.

[6] Shang, L.; Dong, S. and Nienhaus, G. U., Ultra-small fluorescent metal nanoclusters: synthesis and biological applications. *Nano Today* **2011**, *6*(4), 401–418.

[7] Zhao, T.; Zhou, T.; Yao, Q.; Hao, C. and Chen, X., Metal nanoclusters: applications in environmental monitoring and cancer therapy. *Journal of Environmental Science and Health, Part C* **2015**, *33*(2), 168–187.

 [8] Feldmann, C., Polyol-Mediated Synthesis of Nanoscale Functional Materials. *Advanced Functional Materials* **2003**, *13*(2), 101–107.

 [9] Santiago-Gonzalez, B.; Monguzzi, A.; Caputo, M.; Villa, C.; Prato, M.; Santambrogio, C.; Torrente, Y.; Meinardi, F. and Brovelli, S., Metal Nanoclusters with Synergistically Engineered Optical and Buffering Activity of Intracellular Reactive Oxygen Species by Compositional and Supramolecular Design. *Scientific Reports* **2017**, *7*(1), 5976.

[10] Zheng, J.; Nicovich, P. R. and Dickson, R. M., Highly fluorescent noble-metal quantum dots. *Annu. Rev. Phys. Chem.* **2007**, *58*, 409–431.

[11] Diez, I. and Ras, R. H. A., Fluorescent silver nanoclusters. *Nanoscale* **2011**, *3*(5), 1963–1970.

[12] (a) Negishi, Y.; Takasugi, Y.; Sato, S.; Yao, H.; Kimura, K. and Tsukuda, T., Magic-numbered Au n clusters protected by glutathione monolayers (n = 18, 21, 25, 28, 32, 39): isolation and spectroscopic characterization. *Journal of the American Chemical Society* **2004**, *126*(21), 6518–6519; (b) Pyykkö, P., Magic nanoclusters of gold. *Nature Nanotechnology* **2007**, *2*, 273.

[13] Aiken, J. D. and Finke, R. G., A review of modern transition-metal nanoclusters: their synthesis, characterization, and applications in catalysis. *Journal of Molecular Catalysis A: Chemical* **1999**, *145*(1), 1–44.

[14] Yau, S. H.; Varnavski, O.; Gilbertson, J. D.; Chandler, B.; Ramakrishna, G. and Goodson III, T., Ultrafast optical study of small gold monolayer protected clusters: a closer look at emission. *The Journal of Physical Chemistry C* **2010**, *114*(38), 15979–15985.

[15] Zheng, J.; Zhou, C.; Yu, M. and Liu, J., Different sized luminescent gold nanoparticles. *Nanoscale* **2012**, *4*(14), 4073–4083.

[16] Chen, W. and Chen, S., Oxygen electroreduction catalyzed by gold nanoclusters: strong core size effects. *Angewandte Chemie International Edition* **2009**, *48*(24), 4386–4389.

[17] Cathcart, N. and Kitaev, V., Silver nanoclusters: single-stage scaleable synthesis of monodisperse species and their chirooptical properties. *The Journal of Physical Chemistry C* **2010**, *114*(38), 16010–16017.

[18] Zhu, M.; Aikens, C. M.; Hendrich, M. P.; Gupta, R.; Qian, H.; Schatz, G. C. and Jin, R., Reversible switching of magnetism in thiolate-protected Au25 superatoms. *Journal of the American Chemical Society* **2009**, *131*(7), 2490–2492.

[19] Yuan, X.; Setyawati, M. I.; Leong, D. T. and Xie, J., Ultrasmall Ag+-rich nanoclusters as highly efficient nanoreservoirs for bacterial killing. *Nano Research* **2014**, *7*(3), 301–307.

[20] Jin, R., Atomically precise metal nanoclusters: stable sizes and optical properties. *Nanoscale* **2015**, *7*(5), 1549–1565.

[21] Yuan, X.; Luo, Z.; Zhang, Q.; Zhang, X.; Zheng, Y.; Lee, J. Y. and Xie, J., Synthesis of highly fluorescent metal (Ag, Au, Pt, and Cu) nanoclusters by electrostatically induced reversible phase transfer. *Acs Nano* **2011**, *5*(11), 8800–8808.

[22] Yang, X.; Gan, L.; Han, L.; Wang, E. and Wang, J., High-Yield Synthesis of Silver Nanoclusters Protected by DNA Monomers and DFT Prediction of their Photoluminescence Properties. *Angewandte Chemie International Edition* **2013**, *52*(7), 2022–2026.

[23] Chen, J.; Liu, J.; Fang, Z. and Zeng, L., Random dsDNA-templated formation of copper nanoparticles as novel fluorescence probes for label-free lead ions detection. *Chemical Communications* **2012**, *48*(7), 1057–1059.

[24] Tanaka, S. I.; Miyazaki, J.; Tiwari, D. K.; Jin, T. and Inouye, Y., Fluorescent platinum nanoclusters: synthesis, purification, characterization, and application to bioimaging. *Angewandte Chemie* **2011**, *123*(2), 451–455.

[25] Sun, H.-T.; Matsushita, Y.; Sakka, Y.; Shirahata, N.; Tanaka, M.; Katsuya, Y.; Gao, H. and Kobayashi, K., Synchrotron X-ray, photoluminescence, and quantum chemistry studies of bismuth-embedded dehydrated zeolite Y. *Journal of the American Chemical Society* **2012**, *134*(6), 2918–2921.

[26] Molard, Y.; Labbé, C.; Cardin, J. and Cordier, S., Sensitization of Er3+ Infrared Photoluminescence Embedded in a Hybrid Organic-Inorganic Copolymer containing Octahedral Molybdenum Clusters. *Advanced Functional Materials* **2013**, *23*(38), 4821–4825.

[27] Andolina, C. M.; Dewar, A. C.; Smith, A. M.; Marbella, L. E.; Hartmann, M. J. and Millstone, J. E., Photoluminescent gold–copper nanoparticle alloys with composition-tunable near-infrared emission. *Journal of the American Chemical Society* **2013**, *135*(14), 5266–5269.

[28] Robinson III, D. A., Synthesis and Characterization of Metal Nanoclusters Stabilized by Dithiolates. **2011**.

[29] Dhanalakshmi, L.; Udayabhaskararao, T. and Pradeep, T., Conversion of double layer charge-stabilized Ag@ citrate colloids to thiol passivated luminescent quantum clusters. *Chemical Communications* **2012**, *48*(6), 859–861.

[30] Shichibu, Y.; Negishi, Y.; Tsunoyama, H.; Kanehara, M.; Teranishi, T. and Tsukuda, T., Extremely high stability of glutathionate-protected Au25 Clusters against core etching. *Small* **2007**, *3*(5), 835–839.

[31] Xia, Y.; Zhao, X.-M.; Kim, E. and Whitesides, G. M., A selective etching solution for use with patterned self-assembled monolayers of alkanethi-olates on gold. *Chemistry of Materials* **1995**, *7*(12), 2332–2337.

[32] (a) Cui, M.; Zhao, Y. and Song, Q., Synthesis, optical properties and applications of ultra-small luminescent gold nanoclusters. *TrAC Trends in Analytical Chemistry* **2014**, *57*, 73–82; (b) Lu, Y. and Chen, W., Sub-nanometre sized metal clusters: from synthetic challenges to the unique property discoveries. *Chemical Society Reviews* **2012**, *41*(9), 3594–3623.

[33] Guével, X. L., Recent Advances on the Synthesis of Metal Quantum Nanoclusters and Their Application for Bioimaging. *IEEE Journal of Selected Topics in Quantum Electronics* **2014**, *20*(3), 45–56.

[34] (a) Xie, J.; Zheng, Y. and Ying, J. Y., Protein-Directed Synthesis of Highly Fluorescent Gold Nanoclusters. *Journal of the American Chemical Society* **2009**, *131*(3), 888–889; (b) Wang, Y.; Wang, Y.; Zhou, F.; Kim, P. and Xia, Y., Protein-Protected Au Clusters as a New Class of Nanoscale Biosensor for Label-Free Fluorescence Detection of Proteases. *Small* **2012**, *8*(24), 3769–3773.

[35] Chan, P.-H. and Chen, Y.-C., Human Serum Albumin Stabilized Gold Nanoclusters as Selective Luminescent Probes for Staphylococ-cus aureus and Methicillin-Resistant Staphylococcus aureus. *Analytical Chemistry* **2012**, *84*(21), 8952–8956.

[36] Liu, C.-L.; Wu, H.-T.; Hsiao, Y.-H.; Lai, C.-W.; Shih, C.-W.; Peng, Y.-K.; Tang, K.-C.; Chang, H.-W.; Chien, Y.-C.; Hsiao, J.-K.; Cheng, J.-T. and Chou, P.-T., Insulin-Directed Synthesis of Fluorescent Gold Nanoclusters: Preservation of Insulin Bioactivity and Versatility in Cell Imaging. *Angewandte Chemie International Edition* **2011**, *50*(31), 7056–7060.

[37] Wen, F.; Dong, Y.; Feng, L.; Wang, S.; Zhang, S. and Zhang, X., Horseradish Peroxidase Functionalized Fluorescent Gold Nanoclusters for Hydrogen Peroxide Sensing. *Analytical Chemistry* **2011**, *83*(4), 1193–1196.

[38] Kawasaki, H.; Hamaguchi, K.; Osaka, I. and Arakawa, R., ph-Dependent Synthesis of Pepsin-Mediated Gold Nanoclusters with Blue Green and Red Fluorescent Emission. *Advanced Functional Materials* **2011**, *21*(18), 3508–3515.

[39] Xavier, P. L.; Chaudhari, K.; Verma, P. K.; Pal, S. K. and Pradeep, T., Luminescent quantum clusters of gold in transferrin family protein, lactoferrin exhibiting FRET. *Nanoscale* **2010**, *2*(12), 2769–2776.

[40] (a) Wei, H.; Wang, Z.; Yang, L.; Tian, S.; Hou, C. and Lu, Y., Lysozyme-stabilized gold fluorescent cluster: synthesis and application as Hg2+ sensor. *Analyst* **2010**, *135*(6), 1406–1410; (b) Das, J. and Kelley, S. O., Tuning the Bacterial Detection Sensitivity of Nanostructured Microelectrodes. *Analytical Chemistry* **2013**, *85*(15), 7333–7338.

[41] Mathew, A. and Pradeep, T., Noble metal clusters: applications in energy, environment, and biology. *Particle & Particle Systems Characterization* **2014**, *31*(10), 1017–1053.

[42] Chen, L.-Y.; Wang, C.-W.; Yuan, Z. and Chang, H.-T., Fluorescent gold nanoclusters: recent advances in sensing and imaging. *Analytical Chemistry* **2014**, *87*(1), 216–229.

[43] Huang, C. C.; Yang, Z.; Lee, K. H. and Chang, H. T., Synthesis of highly fluorescent gold nanoparticles for sensing mercury (II). *Angewandte Chemie* **2007**, *119*(36), 6948–6952.

[44] Xie, J.; Zheng, Y. and Ying, J. Y., Highly selective and ultrasensitive detection of Hg2+ based on fluorescence quenching of Au nanoclusters by Hg2+–Au+ interactions. *Chemical Communications* **2010**, *46*(6), 961–963.

[45] Lin, Y.-H. and Tseng, W.-L., Ultrasensitive sensing of Hg2+ and CH3Hg+ based on the fluorescence quenching of lysozyme type VI-stabilized gold nanoclusters. *Analytical Chemistry* **2010**, *82*(22), 9194–9200.

[46] Guo, C. and Irudayaraj, J., Fluorescent Ag clusters via a protein-directed approach as a Hg (II) ion sensor. *Analytical Chemistry* **2011**, *83*(8), 2883–2889.

[47] Chen, W.; Tu, X. and Guo, X., Fluorescent gold nanoparticles-based fluorescence sensor for Cu2+ ions. *Chemical Communications* **2009**, *13*, 1736–1738.

[48] Yang, X.; Yang, L.; Dou, Y. and Zhu, S., Synthesis of highly fluorescent lysine-stabilized Au nanoclusters for sensitive and selective detection of Cu 2+ ion. *Journal of Materials Chemistry C* **2013**, *1*(41), 6748–6751.

[49] Shang, L. and Dong, S., Silver nanocluster-based fluorescent sensors for sensitive detection of Cu (II). *Journal of Materials Chemistry* **2008**, *18*(39), 4636–4640.

[50] Wang, Y.-Q.; Zhao, T.; He, X.-W.; Li, W.-Y. and Zhang, Y.-K., A novel core-satellite CdTe/Silica/Au NCs hybrid sphere as dual-emission ratiometric fluorescent probe for Cu2+. *Biosensors and Bioelectronics* **2014**, *51*, 40–46.

[51] Lan, G.-Y.; Huang, C.-C. and Chang, H.-T., Silver nanoclusters as fluorescent probes for selective and sensitive detection of copper ions. *Chemical Communications* **2010**, *46*(8), 1257–1259.

[52] Liu, H.; Zhang, X.; Wu, X.; Jiang, L.; Burda, C. and Zhu, J.-J., Rapid sonochemical synthesis of highly luminescent non-toxic AuNCs and Au@ AgNCs and Cu (II) sensing. *Chemical Communications* **2011**, *47*(14), 4237–4239.

[53] Halawa, M. I.; Wu, F.; Nsabimana, A.; Lou, B. and Xu, G., Inositol directed facile "green" synthesis of fluorescent gold nanoclusters as selective and sensitive detecting probes of ferric ions. *Sensors and Actuators B: Chemical* **2018**, *257*, 980–987.

[54] Huang, Y.; Zhang, H.; Xu, X.; Zhou, J.; Lu, F.; Zhang, Z.; Hu, Z. and Luo, J., Fast synthesis of porous copper nanoclusters for fluorescence detection of iron ions in water samples. *Spectrochimica Acta Part A: Molecular and Biomolecular Spectroscopy* **2018**, *202*, 65–69.

[55] Boonmee, C.; Promarak, V.; Tuntulani, T. and Ngeontae, W., Cysteamine-capped copper nanoclusters as a highly selective turn-on fluorescent assay for the detection of aluminum ions. *Talanta* **2018**, *178*, 796–804.

[56] Hu, X.; Mao, X.; Zhang, X. and Huang, Y., One-step synthesis of orange fluorescent copper nanoclusters for sensitive and selective sensing of Al3+ ions in food samples. *Sensors and Actuators B: Chemical* **2017**, *247*, 312–318.

[57] Lin, L.; Hu, Y.; Zhang, L.; Huang, Y. and Zhao, S., Photoluminescence light-up detection of zinc ion and imaging in living cells based on the aggregation induced emission enhancement of glutathione-capped copper nanoclusters. *Biosensors and Bioelectronics* **2017**, *94*, 523–529.

[58] (a) Liu, W.; Peterson, P. E.; Langston, J. A.; Jin, X.; Zhou, X.; Fisher, A. J. and Toney, M. D., Kinetic and Crystallographic Analysis of Active Site Mutants of Escherichia coli γ-Aminobutyrate Aminotransferase. *Biochemistry* **2005**, *44*(8), 2982–2992; (b) Hayashi, H.; Mizuguchi, H.; Miyahara, I.; Islam, M. M.; Ikushiro, H.; Nakajima, Y.; Hirotsu, K. and Kagamiyama, H., Strain and catalysis in aspartate aminotransferase. *Biochimica et Biophysica Acta (BBA) – Proteins and Proteomics* **2003**, *1647*(1–2), 103–109.

[59] Fogle, E. J.; Liu, W.; Woon, S.-T.; Keller, J. W. and Toney, M. D., Role of Q52 in Catalysis of Decarboxylation and Transamination in Dialkylglycine Decarboxylase. *Biochemistry* **2005**, *44*(50), 16392–16404.

[60] Sun, S. and Toney, M. D., Evidence for a Two-Base Mechanism Involving Tyrosine-265 from Arginine-219 Mutants of Alanine Racemase. *Biochemistry* **1999**, *38*(13), 4058–4065.

[61] Mukherjee, T.; Costa Pessoa, J. O.; Kumar, A. and Sarkar, A. R., Oxidovanadium(IV) Schiff Base Complex Derived from Vitamin B6: Synthesis, Characterization, and Insulin Enhancing Properties. *Inorganic Chemistry* **2011**, *50*(10), 4349–4361.

[62] Schneider, G.; Käck, H. and Lindqvist, Y., The manifold of vitamin B6 dependent enzymes. *Structure* **2000**, *8*(1), R1–R6.

[63] (a) Bothra, S.; Kumar, R. and Sahoo, S. K., Pyridoxal derivative functionalized gold nanoparticles for colorimetric determination of zinc(ii) and aluminium(iii). *RSC Advances* **2015**, *5*(118), 97690–97695; (b) Sharma, D.; Moirangthem, A.; Kumar, R.; Ashok Kumar, S. K.; Kuwar, A.; Callan, J. F.; Basu, A. and Sahoo, S. K., Pyridoxal-thiosemicarbazide: its anion sensing ability and application in living cells imaging. *RSC Advances* **2015**, *5*(63), 50741–50746.

[64] (a) Hu, D.; Sheng, Z.; Gong, P.; Zhang, P. and Cai, L., Highly selective fluorescent sensors for Hg2+ based on bovine serum albumin-capped gold nanoclusters. *Analyst* **2010**, *135*(6), 1411–1416; (b) Huang, C. C.; Yang, Z.; Lee, K. H. and Chang, H. T., Synthesis of highly fluorescent gold nanoparticles for sensing mercury(II). *Angew Chem Int Ed Engl* **2007**, *46*(36), 6824–8.

[65] Bothra, S.; Babu, L. T.; Paira, P.; Ashok Kumar, S.; Kumar, R. and Sahoo, S. K., A biomimetic approach to conjugate vitamin B6 cofactor with the lysozyme cocooned fluorescent AuNCs and its application in turn-on sensing of zinc(II) in environmental and biological samples. *Analytical and Bioanalytical Chemistry* **2017**, *410*(1), 201–210.

[66] Bothra, S.; Paira, P.; Kumar S. K. A.; Kumar, R. and Sahoo, S. K., Vitamin B6 Cofactor-Conjugated Polyethyleneimine-Passivated Silver Nanoclusters for Fluorescent Sensing of Zn2+ and Cd2+ Using Chemically Modified Cellulose Strips. *ChemistrySelect* **2017**, *2*(21), 6023–6029.

15

Solid Phase Astrochemistry: Is Space a Realistic Laboratory for Life's Molecules?

Sohan Jheeta[1,]* and Elias Chatzitheodoridis[2]

[1]Network of Researchers on the Chemical Evolution of Life (NoR CEL), UK
[2]National Technical University of Athens (NTUA), Greece
E-mail: sohan@sohanjheeta.com
*Corresponding Author

The beguiling question of the origin of life is the result of a gradual chemical evolution of molecules from simpler initial molecules; the chemical evolution, in essence, led to what is often referred to as a super-hypercomplex chemistry of which the final end-point is life. In order to understand the ultimate emergence of life we need to ask a preliminary question: where and how these initial molecules were formed? There are several places where the molecules could be made including in atmospheric lightning; at the ebb and flow of tidal cycles; at hydrothermal vents on the sea floor; in-between dense and less dense atmospheric layers; and further afield in the vastness of the huge dark molecular clouds where the Solar System was made. In this respect the dark molecular clouds act as huge chemical processing "factory". This chapter briefly aims to cover this area of astrochemistry as well as touching upon the delivery of these initial molecules to surface of the Earth. Whilst this chapter is written in the context of life on Earth, it should be noted that the same chemistry is applicable to any dark molecular cloud, any solar system and to all the planets within a proto-planetary disk.

15.1 Introduction

Planet Earth has one particular unique attribute, i.e., it harbors life. This attribute immediately invokes a very important question: how did this life

189

originate? The answer to this question is not straightforward as it poses yet another: where were the molecules that initiated the formation of life made? Both of these questions are interlinked in that the discussion around the latter shows the way to the answer posed in the former. The process of creating life from simple, life friendly organic (e.g. formaldehyde) molecules is, in reality, seamless and can be termed chemical evolution. In this regard, it is pertinent to note that the former question involves areas of astrobiology and the latter focuses on astrochemistry. (Readers may wish to consult the paper entitled, "Final frontiers: the hunt for life elsewhere in the Universe" by [1], in which the origin of life in general is also discussed). In this chapter, we principally address the areas of astrochemistry.

15.1.1 The Molecules of Life

It is now well established that the molecules of life can originate at numerous places including at the boundaries between dense and less dense atmospheric layers, such as in the atmosphere of Saturn's moon, Titan [2]; during lightning strikes in the primordial planetary atmosphere, as shown via the electric discharge experiments carried out by Stanley Miller [3] (Table 15.1); within small pools of water at the base of volcanoes [4]; during the cyclic ebb and flow of tides on the seashores [5]; in alkaline hydrothermal vents on the ancient sea floors [6]; via impactors, namely meteorites, comets, and asteroids (Table 15.1, [7, 8]); and, of course, by astrochemical processes in space.

Astrochemistry is the study of the formation of molecules under simulated space conditions. To date over 190 molecules have been detected in dark molecular clouds, technically known as the interstellar medium (ISM), [9], such as the Horsehead Nebula (Figure 15.1). The molecules could be as simple as monatomic, e.g., helium (He) and homoatomic molecules consisting of atoms of the same element, (diatomic) as in dihydrogen (H_2), or heteroatomic diatoms such as carbon monoxide (CO). Other molecules could be polyatomic, including water (H_2O), hydrogen cyanide (HCN), formyl radical (HCO), isocyanic acid (HNCO), as well as even more complex ones like cyanoacetylene (C_3HN, 5 atoms), methenamine (CH_3N, 6 atoms), acetaldehyde (CH_3CHO, 7 atoms), and even molecules with 13 atoms (cyanopentaacetylene with mass of 147). Included in the inventory are also exotic molecules, such as isotopes deuterium (D), ^{13}C, ^{17}O, ^{18}O; radicals OH, CN, CH, CH^+; metals like aluminum monochloride (AlCl) and nonmetals, e.g, carbonyl sulfide (OCS), as well as fullerenes with 70 carbon atoms (C_{70}) and a mass of 840. There are also much postulated constituents of the ISM

Table 15.1 Comparison of amino acids detected via the Urey–Miller experiment with those found in the Murchison meteorite [adapted from [7]].

Amino Acids	Synthesized in the Miller–Urey Experiments	Found in the Murchison Meteorite
Glycine	✓	✓
Alanine	✓	✓
α-amino-N-butyric acid	✓	✓
α-aminoisobutyric acid	✓	✓
Valine	✓	✓
Norvaline	✓	✓
Isovaline	✓	✓
Proline	✓	✓
Pipecolic acid	✓	✓
Aspartic acid	✓	✓
Glutamic acid	✓	✓
β-alanine	✓	✓
β-amino-N-butyric acid	✓	✓
β-aminoisobutyric acid	✓	✓
Υ-aminobutyric acid	✓	✓
Sarcosine	✓	✓
N-ethylglycine	✓	✓
N-methylalanine	✓	✓

and circumstellar disks, i.e., the polycyclic aromatic hydrocarbons which are basically a myriad of hydrocarbon rings [10].

15.1.2 Evidence for Molecules in Space

The presence of molecules in space can be ascertained as follows: via detection of signals from far-flung ISMs by radio-telescopes; isolation of organic molecules in carbonaceous meteorites (e.g., Murchison meteorite); return samples from unmanned missions to comets (Stardust spacecraft to Wild 2 comet, 1999–2006) and asteroids (e.g., dwarf planets such as Ceres, Dawn mission, 2007–2018). In addition, laboratory simulation experiments validate the possibility of molecule formation in space conditions. A short appraisal of these areas is as follows:

15.1.2.1 Radio-telescope

In simple terms electrons within molecules absorb the electromagnetic radiation emitted (generally infrared [IR] light) by a nearby star; when electrons relax to their ground state, they reemit IR signals at a specific wavelength

Figure 15.1   The best known and studied dark molecular cloud is the Horsehead nebula (photo courtesy of the NASA). These molecular clouds are dark because they contain 80% H_2, 19% He, and $\sim$1% other elements and molecules. They also contain 1 interstellar dust particle (IDP) for 10^6 H_2 giving them the appearance of being black. This image's appearance is burgundy because it is taken through an infrared (IR) filter [11].

for given molecules – for example, CO_2 has the following wavenumbers 660, 2347, 3602, and 3708 cm^{-1} (Figure 15.2).

The limiting factors for radio-telescopic detection of IR signals are the immense distances involved, background radiation, Earth's atmospheric conditions and working conditions pertaining to instrumentation. These reduce the signal/noise ratio arising from distant ISMs to indecipherable levels thereby impeding identification of relevant molecules [1, 14, 15].

15.1.2.2 Carbonaceous meteorites

Impactors, including meteorites, were made during the formation of planetary disks and contain an inventory of organic materials. The analysis of these impactors yielded a wealth of information regarding the organics they contain, including carboxylic and hydrocarboxylic acids, alcohols, aliphatic, and aromatic hydrocarbons, amino acids, sugars, and nitrogen bases [16–18] Some of the molecules shown in Table 15.2 belong to the glycolytic pathway

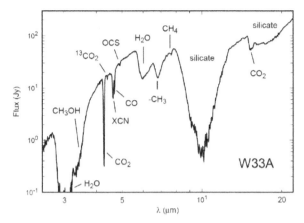

Figure 15.2 A spectrum of the protostar W33A. Dominant absorption bands are visible at 3.0 μm (H_2O ice), 4.27 μm (CO_2 ice), 4.67 μm (CO ice), and 10 μm (silicates). The spectrum between 2 and 25 μm shows several additional species, many of them organic in nature; the respective wavenumbers for the frequencies mentioned are 3333, 2342, 2141, and 1000 cm^{-1} [12, 13].

(e.g., pyruvic acid) or are part of the tricarboxylic acid cycle (eg citric acid), both of which are used in metabolic pathways.

15.1.2.3 Return sample by unmanned missions
In 2004, the Stardust Spacecraft Mission rendezvoused with the comet 82P/Wild 2. The material being evaporated by solar radiation formed a tail (coma) behind the comet as it hurtled toward the Sun. The emitted material was then collected as it impacted the aerogel contained in the petri dish on board the Stardust Spacecraft, which then successfully brought back the collected sample to Earth. Upon an analysis of these IDPs, the nonchiral amino acid, glycine, was detected [20–22]; glycine being the first amino acid in a series of 20 used in the biology on Earth. Other molecules detected in the coma of comets are as indicated in Table 15.3.

15.1.2.4 Laboratory simulation experiments
It is possible to recreate space conditions in laboratory settings in which experiments can be performed in order to create chemical compounds. These conditions consist of extremely low pressures (e.g., 10^{-10} mb) obtained using a series of ultrahigh vacuum pumps, as well as low temperatures such as those prevalent in the ISM (30–50 K), which are generated using liquid helium.

Table 15.2   Compounds identified in the Murchison, Murray, and Alan Hills (ALH) 83102 carbonaceous meteorites using a gas chromatography-mass spectrometer. It should be noted that as the complexity of molecule increases, due to the presence of functional groups, the IR spectra also becomes more difficult to decipher, e.g., pyruvic acid in the Table below is an alpha-keto acid, which means it contains three functional groups, namely methyl, kito and carboxylic, compared to, say, methane which has none. [Adapted from a paper by [19]]. Readers may wish to consult this article for further in-depth information.

Some of the metabolically important compounds identified in meteorites
List of keto acids
Pyruvic acid
Acetoacetic acid
Levulinic acid
3-methyl-4-oxopentanoic acid
5-oxohexanoic acid
6-oxoheptanoic acid
7-oxooctanoic acid
α-keto glutaric acid
4-hydroxy-4methyl-2-ketoglutaric acid (pyruvic acid dimer)
List of hydroxy tricarboxylic acids
1-hydroxy-1,1,2-ethanettricarboxylic acid
Citric acid
Isocitric acid
List of tricarboxylic acid
1,2,3,4-propanetricarboxylic acid

This results in the formation of astrophysical ices on a crystal target, e.g., zinc selenide; such ices are then irradiated with either particle (e.g., e^-, H, H^+, D^+, He, He^+, etc.) or electromagnetic (e.g., UV light) radiation to form new chemical compounds. The newly made compounds are then identified using Fourier-transform infrared spectroscopy (Figure 15.3).

The types of products obtained are shown in Table 15.3 [1] and are compared with results obtained by [24] as well as organics in the comas of comets [25], indicating that space is a realistic laboratory for the synthesis of life's molecules.

The outcome of such investigations confirms the presence of organic molecules just about everywhere in the Universe and supports the idea that these molecules were delivered on to the Earth via impactors during the heavy bombardment period around 4 billion (10^9) years ago [8].

Table 15.3 The experimental results of Jheeta et al. [26] and Gerakines et al. [24] show that organic molecules can be made by irradiating CH_3OH or 1:1 mixture of $NH_3:CH_4OH$ ice at 30 K with 1 keV (at 10 µA) electrons (radiolysis) or irradiated with UV light (photolysis) in simulated space conditions. Such energies break the chemical bonds within the ice forming radicals which then rearrange themselves to form new compounds. Ferris [25] reported that the organic molecules contained in the coma of comets originated in the ISM. The table also shows the presence of volatiles within the IDP mantle [27].

	Radiolysis, via 1 keV (10 µA) e⁻ of Methanol [1]	Radiolysis via 1 keV (10 µA) e⁻ of 1:1 $NH_3:CH_4OH$ [1]	Photolysis via UV Light of Methanol [24]	Organics in the Comas of Comets [25]	Precursor Molecules in the IDP Mantles* [27]
Methylformate	✓	✓	✓	✓	
Methane	✓	✓	✓	✓	✓
Hydroxymethyl	✓	✓	✓		
Formic acid	✓	✓		✓	✓
Formaldehyde	✓	✓	✓		✓
Formyl radical	✓	✓	✓		
Carbon monoxide	✓	✓	✓		✓
Carbon dioxide	✓	✓	✓		✓
Ethanol			✓		
Formamide		✓		✓	
Methanol				✓	✓
Ethylene				✓	
Methylacetylene				✓	
Acetonitrile				✓	
Acetylene				✓	
Ethane				✓	
Hydrogen cyanide				✓	
Cyanic acid		✓			
Cyanate		✓			
Hydrogen sulfide					✓
Carbonyl sulfide					✓
Ammonia					✓
Dihydrogen					✓

*Interstellar dust particles (IDPs) would have a watery ice mantle because after H_2 and He, the next most abundant molecule in space is H_2O.

15.1.3 Interstellar Medium as Chemical Factories

The processes by which precursor molecules such as ammonia (NH_3), methanol (CH_3OH) methane (CH_4), CO, and CO_2 (Table 15.3, column 6) collide with both one another, as well as the IDPs in order to bring about reactions is not yet fully understood [28]. Since the pressures in the ISM

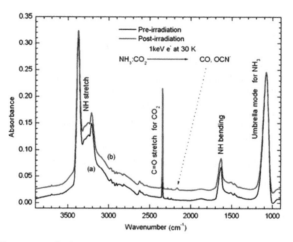

Figure 15.3 Shows a typical spectrum obtained during the irradiation of 1:1 mixture of ammonia and carbon dioxide ice at 30 K using 1 keV (at 10 μA) electrons. The black trace is for preirradiation and the red trace depicts postirradiation at ∼15 min [23].

are so incredibly low (down to 10^{-13} mb), the precursor molecules and the IDPs are really far apart; for example, the number of molecules in the air on Earth is in the order of $10^{19} cm^{-3}$ compared to 10^6 cm^{-3} in the ISMs, therefore the likelihood of two or more molecules colliding is lower in the ISMs. In order to get a handle on space chemistry some knowledge of *collision theory* is necessary. Essentially, the theory demands that molecules must collide; further, the collision must have the correct orientation for the reaction to proceed; such collisions must have sufficient energy. These three requirements are encompassed in Arrhenius' equation (Equation 15.1):

$$k = Ae^{-Ea/(RT)} \tag{15.1}$$

where k is the reaction rate constant and A is the pre-exponential factor, a constant for each chemical reaction. According to the collision theory, A is the frequency of collisions with the correct orientation and E_a is the activation of energy for a given reaction and has the same units as RT. So, the lower the E_a of a reaction, the higher the k for the given correctly oriented collisions, and T is the absolute temperature (K). Like E_a, the temperature also affects the reaction rate. A fall in temperature would reduce the value of k for the same correct collisions rate as discussed earlier, and R is the universal gas constant (8.3145 J Mol^{-1} K^{-1}).

Given this information, the ISMs were, metaphorically speaking, huge laboratories and more prolific when compared to the other methods of forming organic molecules. But this does not necessarily mean that space would have provided all the necessary molecules for life to have emerged on Earth as the other methods of synthesis could also have been required.

The precursor molecules within the ISM are processed via the general reactions as shown in Table 15.4. These reactions could take place either in the gas phase, which makes up 75% of the reactions, or they could take place on the surface of IDPs (25%).

Based on the literature, IDPs (Figures 15.4a and 15.4b) are small-sized (0.1 μm) particles which are present within the ISM (e.g., Barnard 68 nebula). There is 1 IDP for every 10^6 molecules of dihydrogen in the ISM. In addition,

Table 15.4 Various types of reactions taking place within a typical ISM. Readers may wish to consult the book, "Interstellar Medium" by Tielens [29] for further information. These reactions could occur between molecule–molecule in the gas phase, or on the surface of the IDPs.

Types of Reactions	General Reactions	Examples of Types of Reactions
Photodissociation	$AB + h\nu \rightarrow A + B$	$N_2 + 121.6$ nm $\rightarrow N + N$ (homolytic free radicals) $H_2O + 280$ nm $\rightarrow H^+ + OH^-$ (heterolytic ions)
Neutral–neutral radicles	$A + B \rightarrow C + D$	$CO + H \rightarrow HCO$ (formyl radical) $NH_2 + CO \rightarrow H + HNCO$ (cyanic acid formation)
Ion–molecule	$A^+ + B \rightarrow C^+ + B$	$Ne^+ + CO \rightarrow Ne + O + C^+$ (carbon ion formation)
Charge-transfer	$A^+ + B \rightarrow A + B^+$	$Ar^+ + N_2 \leftrightarrow N_2^+$
Radiative association	$A + B \rightarrow AB + h\nu$	$CH_3^+ + H_2O \rightarrow CH_3H_2O^+ + h\nu$ (methanol formation intermediate – see the following reactions)
Dissociative recombination	$A^+ + e^- \rightarrow C + D$	$CH_3H_2O^+ + e^- \rightarrow H + CH_3OH$ (methanol formation) $N_4C_2N^+ + e^- \rightarrow H + CH_3CN$ (acetonitrile formation)
Collisional association*	$A + B + M \rightarrow AB + M$	$H + H + M \rightarrow H_2 + M$ (H_2 formation) $O + O_2 + M \rightarrow O_3 + M$ (Chapman reaction)
Associative detachment	$A^- + B \rightarrow AB + e^-$	$H^- + H \rightarrow H_2 + e^-$

*Three body reactions, where M being the third body which absorbs the liberated energy during the reaction.
Also, the only known way of making H_2.

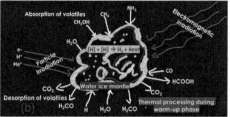

Figure 15.4   (a) An actual IDP brought back to the Earth by the Stardust spacecraft in 2006 and (b) is a diagrammatical representation of the IDP showing impinging particle and electromagnetic radiations, processes of absorption and desorption of various molecules, as well as thermal processing during the warm-up phase. Also note that during the formation of dihydrogen, the IDP acts as the third body for absorbing the excess heat being produced [31].

IDPs are composed mainly of silicates (80%) and carbonaceous material (20%) and are amorphous and porous; the resulting texture gives IDPs an increased surface area upon which forms an icy mantle made of frozen water containing both volatile organic and inorganic compounds (Table 15.3, column 6). These particles are made within the asymptotic giant branch stars themselves. Scientists have confirmed that the ejecta (10^8–10^9 M$_\odot$) of Cassiopeia A, a supernova remnant, laying almost at the edge of the Universe at 13 billion light years from Earth, produced enough IDP material to make more than 10,000 Earths. The ejecta being composed mainly of tiny particles of protosilicates, silicon dioxide, iron oxide, pyroxene, carbon, and aluminum oxide [30]. Thus, IDPs are ubiquitous and thus are vitally important for the chemical evolution of the Universe.

In addition, precursor molecules could also be created via thermal processing, that is, when IDPs become heated, due to shockwaves generated by a local supernova, new molecules may be formed as illustrated in Figure 15.5.

It has also been shown that the essential chiral molecules for life (e.g., sugars, and in particular ribose, a prerequisite for RNA) can be made in the ISM [20, 33], although IR signals for these still remain difficult to decipher, noting that sugars, amino acids, and nitrogen bases (e.g., uracil and thymine) have been isolated from meteorites [18].

15.1.4 Delivery of Molecules

During the early period of the Earth's history, ~4–4.3 billion years ago, the Solar System was, by and large, a very violent place, known as the heavy bombardment era. As mentioned earlier, impactors played a pivotal role in

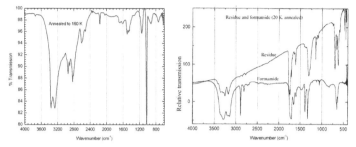

Figure 15.5 (a) Formamide ($HCONH_2$) was formed (at 160 K) when the irradiated ice of NH_3:CH_3OH was annealed at steps of $10°C$ [26]. and (b) This was compared with the work of Khanna et al. [32] who first annealed the sample to 165 K and then recooled to 20 K.

the delivery of organic molecules, water, and early primordial atmospheric gases, for example, CH_4 onto the Earth. Evidence of this violent period is available courtesy of the Moon and Mercury as both of these bodies do not have any discernible atmosphere and therefore their impact craters are as they would have been originally, since they have not been subject to any withering processes. It is estimated that enough organic materials were delivered onto the surface of the Earth during this period to form a layer of organic material between a half and one metre thick which covered the entire globe [7]. This would have included the organic molecules identified in Tables 15.1–15.3. For example, exoplanets in the Trappist-1 (2MASS J23062928-0502285) system could be a likely candidate for being laden with a huge amount of organic materials as well as water, making the possibility of life taking sometime in the future a real prospect [34].

15.2 Conclusion

It is clear that molecules could be formed in space although to date only around 190 have been identified. This is because the many aspects of molecules (e.g., functional groups such as carboxylic acid) make the majority indecipherable. It is believed that these molecules were delivered on to the surface of the Earth by impactors. Even if these figures (a half and one metre thick) were substantially overestimated, suffice to say that enough material could be made in space and delivered on to the surface of any other planet within any planetary disk anywhere in the Cosmos, noting that planetary disks are made within the ISM.

References

[1] Jheeta, S. (2013a). "Final frontiers: the hunt for life elsewhere in the Universe". *Astrophysics and Space Science*, **348**(1): 1–10. DOI 10.1007/s10509-013-1536-9.

[2] Laine, PE. and Jheeta, S. (2017). "Astrobiology: Exploring Life on Earth and Beyond. Habitability of the Universe before Earth". Academic Press Publication. Chapter 14, Pages: 321–342.

[3] Miller, SL. (1953). "A production of amino acids under possible primitive Earth conditions". *Science*, **117**(3046): 528–529. DOI: http://www.jstor.org/stable/1680569.

[4] Jheeta, S. (2017). "The landscape of the emergence of life". *Life*, **7**(27): 1–11.

[5] Lathe, R. (2004). "Fast tidal cycling and the origin of life". *Icarus*, **168**: 18–22.

[6] Russell, MJ., Barge, LM., Bhartia, R., Bocanegra, D., Bracher, PJ., Branscomb, E. et al. (2014). "The drive to life on wet and icy worlds". *Astrobiology*, **14**: 308–343.

[7] Gilmour, I. and Wright, I. (Eds). (2003). The window of opportunity. In "Origins of Earth and Life". MRM Graphics, Winslow, UK. pp. 100–102.

[8] Chyba, CF., Thomas, PJ., Brookshaw, L., Sagan, C. et al. (1990). "Cometary delivery of organic molecules to the early Earth". *Science*, **249**: 336–373.

[9] https://en.wikipedia.org/wiki/List_of_interstellar_and_circumstellar_molecules.

[10] Lovas, FJ., McMahon, RJ., Grabow, J-U., Schnell, M., Mack, J., Scott, LT. et al. (2005). "Interstellar chemistry: a strategy for detecting polycyclic promatic hydrocarbons in space". *JACS*, **127**(12): 4345–434.

[11] https://www.nasa.gov/multimedia/imagegallery/image_feature_2493.html.

[12] Gibb, E., Whittet, DCB., Schutte, WA., Chiar, J., Ehrenfreund, P. et al. (2000). "An inventory of interstellar ices toward the imbedded protostar W33A". *The Astrophysical Journal*, **536**: 347–356.

[13] Ehrenfreund, P. and Charnley, SB. (2000). "Organic molecules in the interstellar medium, comets and meteorites: a voyage from dark clouds to the early Earth". *Annual Review of Astronomy and Astrophysics*, **38**: 427–483.

[14] Hudson, RL. (2006). "Astrochemistry examples in the classroom". *Journal of Chemical Education*, **83**: 1611–1616.

[15] Onaka, T., Matsumoto, H., Sakon, I., Kaneda, H. et al. (2008). "Organic compounds in galaxies". *Proceedings IAU Symposium*, **251**: 229–236.

[16] Cronin, JR. and Pizzarello, S. (1986). "Amino acids of the Murchison meteorite. III. Seven carbon acyclic primary alpha-amino alkanoic acids". *Cosmochimica Acta*, **50**: 2419–2427.

[17] Cronin, JR. (1989). "Origin of organic compounds in carbonaceous chondrites". *Advances in Space Research*, **9**: 59–65.

[18] Sephton, MA. (2002). "Organic compounds in carbonaceous meteorites". *Natural Product Reports*, **19**: 292–311.

[19] Cooper, G., Reed, C., Nguyen, D., Carter, M., Wang, Y. et al. (2011). "Detection and formation scenario of citric acid, pyruvic acid, and other possible metabolism precursors in carbonaceous meteorites". *PNAS*, **108**(34): 14015–14020.

[20] Elsila, JE., Glavin, DP. and Dworkin, JP. (2009). "Cometary glycine detected in samples returned by Stardust". *Meteoritics and Planetary Science*, **44**(9): 1323–1330.

[21] Kwok, S. (2016). "Complex organics in space from solar system to distant galaxies". *Astronomy and Astrophysics Reviews*, **24**: 1–27.

[22] Sandford, SA., Engrand, C. and Rotundi, A. (2016). "Organic matter in cosmic dust". *Elements*, **12**: 185–189.

[23] Jheeta, S., Ptasinska, S., Sivaraman, B., Mason, NJ. et al. (2012). "The irradiation of 1:1 mixture of ammonia:carbon dioxide ice at 30 K using 1 keV electrons". *Chemical Physics Letters*, **543**: 208–212.

[24] Gerakines, PA., Schutte, WA. and Ehrenfreund, P. (1996). "Ultraviolet processing of interstellar ice analogs. 1. Pure ices". *Astronomy and Astrophysics*, **312**(1): 289–305.

[25] Ferris, JP. (2006). "Montmorillonite-catalzed formation of RNA oligomers: the possible role of catalysis in the origin of life". *Philosophical Transactions of the Royal Society B*, **361**: 1777–1786.

[26] Jheeta, S., Domaracka, A., Ptasinska, S., Sivaraman, B., Mason, NJ. et al. (2013b). "The irradiation of pure CH_3OH and 1:1 mixture of NH_3:CH_3OH ices at 30 K using low energy electrons". *Chemical Physics Letters*, **556**: 359–364.

[27] Irvine, WM. (1998). "Extraterrestrial organic matter: A review". *Origins of Life and Evolution of Biospheres*, **28**(4–6): 365–383.

[28] Herbst, E. (2001). "The chemistry of interstellar space". *Chemical Society Reviews*, **30**: 168–176.

[29] Tielens, AGGM. (2006). "Interstellar Medium". Cambridge University Press, UK.

[30] Rho, J., Kozasa, T., Reach, WT., Smith, JD., Rudnick, L., DeLaney T J. et al. (2008). "Freshly formed dust in the Cassiopeia a supernova remnant as revealed by the Spitzer space telescope." *The Astrophysical Journal*, **673**(1): 271–282.

[31] https://en.wikipedia.org/wiki/Cosmic_dust.

[32] Khanna, RK., Lowenthal, MS., Ammon, HL., Moore, MH. et al. (2002). "Molecular structure and infrared spectrum of solid amino formate (HCO_2NH_2): Relevance to interstellar ices." *The Astrophysical Journal Supplement Series*, **140**(2): 457–464.

[33] Saladino, R., Crestini, C., Ciciriello, F., Costanzo, G., Di Mauro, E. et al. (2007). "Formamide chemistry and the origin of informational polymers". *Chemistry and Biodiversity*, **4**(4): 694–720.

[34] https://www.eso.org/public/news/eso1805.

16

Synthesis and Characterization of Hydroxyapatite Nanoparticles Using a Novel Combustion Technique for Bone Tissue Engineering

**Yesoda Velukutty Swapna, Jijimon K. Thomas[*],
Christopher Thresiamm Mathew, Jayachandran Santhakumari
Lakshmi, Steffy Maria Jose and Sam Solomon**

Department of Physics, Mar Ivanios College, Thiruvananthapuram 695015, Kerala, India
E-mail: jkthomasemrl@yahoo.com
[*]Corresponding Author

Hydroxyapatite (HA) is one of the most demanding bioceramic materials which have applications in bone tissue engineering. Synthesis of nanoparticles of HA having very low particle size by a novel single step autoigniting combustion technique for the first time is presented in this paper. The combustion-synthesized powder was characterized by different powder characterization techniques. The X-ray diffraction pattern reveals that the as-synthesized powder is phase pure with an average crystallite size of $\sim$25 nm. The Fourier-transform infrared spectrum confirms their phase purity. The ultraviolet (UV)-visible spectroscopy reveals that the material absorbs heavily in the UV region. The band gap of the as-prepared sample measured using the Kubelka–Munk method is found to be 5.4 eV. The thermogravimetric analysis discloses its thermal stability even at 800°C which is favorable for its applications in the synthesis of nanoscaffolds. The results clearly show that the high quality nanostructured HA synthesized by an economic autoigniting combustion technique can effectively be used in futuristic scaffolds for tissue regeneration.

16.1 Introduction

Synthetic hydroxyapatite (HA) has been used extensively in different medical applications as biomaterials due to its excellent biocompatibility with human tissues [1–4]. Due to its close chemical similarity to the natural form found in the inorganic component of the bone matrix, there have been extensive research efforts into employing synthetic HA as a bone substitute in a number of clinical applications such as bone augmentation, coating metal implants, and filling components in both bone and teeth [5–9]. It is possible to improve the mechanical properties of the HA ceramic by controlling important parameters of powder precursors such as particle size and shape, and particle distribution and agglomeration [21]. Nanocrystalline HA powders exhibit greater surface area [10]. It can provide improved sinterability and enhanced densification to reduce sintering temperature, which could improve the mechanical properties of the HA ceramic [11]. Moreover, nanometer-sized HA is also expected to have better bioactivity than coarser crystals [12, 13]. Nanophase ceramics clearly represent a unique and promising class of orthopedic/dental implant formulations with improved osseointegrative properties [13, 14]. Therefore, scaffolds fabricated from high quality nanocrystalline HA powders are expected to have better mechanical properties and improved osseointegrative properties.

In the present work, a novel combustion method is employed to effectively synthesize phase-pure nanocrystalline HA. The as-synthesized powder was characterized for structural determination, particle size, thermal stability, and optical behavior.

16.2 Methods

High quality HA ceramic powder was prepared (Ca/P molar ratio: 1.67) using calcium nitrate tetrahydrate Ca $(NO_3)_2 \cdot 4H_2O$ and di-Ammonium hydrogen phosphate $(NH_4)_2HPO_4$ using a single step autoigniting combustion technique developed by us. The precisely weighed metal salts were dissolved in double distilled water. To the precursor solution citric acid was added which acts both as a fuel and complexing agent. The quantity of citric acid was calculated for maximum release of energy during combustion [15]. The oxidant-to-fuel ratio of the system was adjusted by adding nitric acid and ammonium hydroxide, and the pH of the solution is maintained acidic. Ammonium nitrate thus formed from the reaction between nitric acid and ammonium hydroxide acts as an extra oxidant without changing the proportion of other reactants in the solution. Ammonium nitrate hence produces an increase in the combustion gas and consequently increases the surface area of the produced powder resulting in the generation of a fine quality nanopowder. The precursor mixture was then placed on a hot plate in a ventilated fume hood. The solution boils on heating and undergoes dehydration accompanied by foam. The foam then ignites by itself on persistent heating giving voluminous and fluffy product of combustion.

The samples were characterized by X-ray diffractometer (Bruker D8 Advance) with Cu Kα radiation in the range of 20–70°C in increments of 0.02°C for the determination of crystalline structure and phase of the nanomaterials. Phase transition in the as-prepared sample was studied using differential thermal analysis (DTA) and thermogravimetric analysis (TGA) by (PerkinElmer, Diamond TG/DTA) thermal analyzer in the range of 50–800°C at a heating rate of 10°C min^{-1} in nitrogen atmosphere. The absorption spectrum of as-prepared samples of HA nanoparticles was recorded using an ultraviolet–visible (UV–Vis) spectrophotometer (UV-1700, Shimadzu, Singapore). Additional information on the phase purity and the presence of any inorganic impurity were obtained using Fourier Transform Infrared (FT-IR) spectrometer (Spectrum 2, PerkinElmer, Singapore) in the range of 400–4000 cm^{-1} using Attenuated Total Reflectance method.

16.3 Results and Discussion

X-ray diffraction (XRD) pattern analysis was used to identify the purity and crystalline size of the synthesized nanometer scale HA powder and the particle size. Analysis of the XRD pattern as shown in Figure 16.1 revealed the presence of a crystalline nanometer scale HA phase. These phases were found to be consistent with the phase reported in the ICDD 740565 database, with the main (h, k, and l) indices for nano-HA: (002), (211), (300), (202), (130), (002), (222), and (213). The crystalline size of the ultrafine HA powder was calculated from the XRD pattern using the method of Debye–Scherrer $D = \frac{K\lambda}{\beta \cos \theta}$ [16, 17] where λ is the wavelength of the monochromatic X-ray beam, β is the full width at half maximum of the peak at the maximum intensity, θ_{hkl} is the peak diffraction angle that satisfies Bragg's law for the (h, k, and l) planes and D is the crystallite size. The crystalline size calculated from the highest reflection peak is 22 nm.

The FT-IR spectrum of the synthesized ultrafine powder is presented in Figure 16.2 and reveals the presence of several bands associated with the nanometer-scale HA. Starting from the right-hand side of the spectrum, the

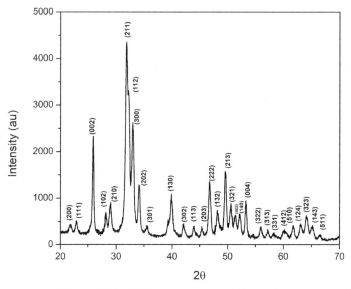

Figure 16.1   XRD pattern of nano-HA powder.

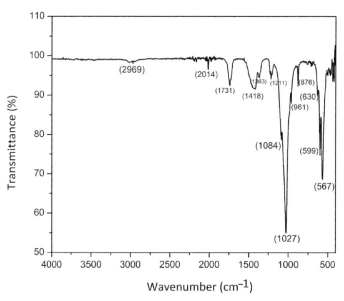

Figure 16.2   FT-IR transmittance spectrum of nano-HA powder.

first two bands we observe are at 567 cm^{-1} and 599 cm^{-1}, which are the results of ν_4 vibrations being produced by the O–P–O mode. The next two bands located at 725 cm^{-1} and 876 cm^{-1} are associated with the carbonate group and clearly indicate the presence of carbonates in the samples. The 961 cm^{-1} band results from the ν_1 symmetric stretching vibrations of the P–O mode. While the very strong peaks located at 1027 cm^{-1} and 1084 cm^{-1} correspond to the PO$_4^{3-}$ functional group (P–O mode) and the weaker peak at 1418 cm^{-1} corresponds to the CO$_3^{2-}$ functional group. The formation of carbonate most likely results from the interaction between atmospheric carbon dioxide and the alkaline HA precursor solution during the synthesis process [18–20]. The peak located at 1731 cm^{-1} also corresponds to a CO$_3^{2-}$ group. The results of the FT-IR analysis clearly indicate that the synthesized ultrafine powders are HA.

The UV–visible spectrum for nano-HA was taken and the absorption wavelength is obtained at 252 nm (Figure 16.3). The band gap energy is calculated as 5.4 eV from the Tauc's plot method using Kubelka–Munk function (Figure 16.4). In the present work we get a single phase nano-HA and the literature review clearly indicates that the band gap can be reduced to 3.95 eV for biphasic HA [21, 22]. This is due to the presence of oxygen vacancies in the biphasic HA lattice on UV irradiation. Thus HA, in the biphasic state, has greater application for environmental remediation.

The analysis of the DTA thermograph presented in Figure 16.5 reveals that there were no exothermic or endothermic reactions taking place over

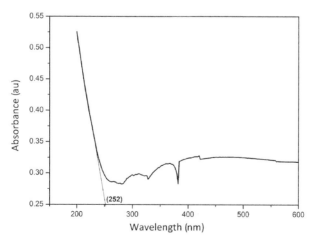

Figure 16.3 UV–Visible absorbance spectrum of nano-HA powder.

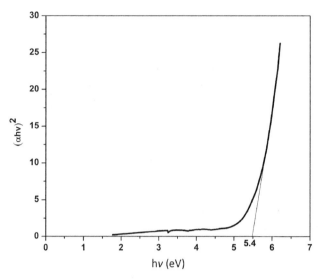

Figure 16.4   Tauc's plot of nano-HA powder.

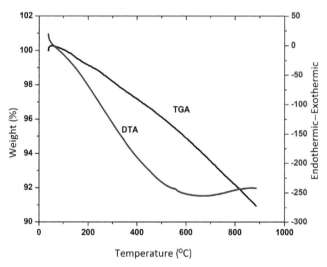

Figure 16.5   TGA and DTA thermographs of the synthesized nano-HA powder.

the 38–880°C temperature range. Furthermore, no phase transformations occurred over this temperature range, which confirmed the physical and chemical stability of the HAP powder sample [23]. Figure 16.5 presents a typical TGA thermograph of a powder sample and reveals a decreasing trend in sample mass due to water loss. The initial sample mass was 100.03 mg and the final mass at the end of the heating cycle was 91.03 mg indicating a water loss of 9 mg (8.99%). Analysis of the thermograph revealed a water loss over the temperature range. The mass loss observed can be due to the evaporation of surface moisture and absorbed water due to the loss of lattice water [24]. From 558°C onwards, there was no further mass loss from the sample indicating the high thermal stability of the HAP powder.

The detailed powder characterization reveals that the HA powder synthesized by a modified combustion show better particulate properties and can be utilized for fabrication of high quality futuristic scaffold materials.

16.4 Conclusions

Synthesis of nanoparticles of HA bioceramics of 22 nm by a modified single step autoigniting combustion technique and its characterization are presented in this paper. The XRD analysis reveals that the sample is phase pure and is crystallized in hexagonal structure with the space group $P6_3/m$. The FT-IR spectroscopy also confirms the phase purity of the sample. The UV–visible spectroscopy reveals that the band gap of the sample is >5 eV. The TGA and DTA curves show that the sample is thermally stable at elevated temperatures. The present study authenticates the efficacy of modified combustion technique to produce phase pure nanostructured HA.

Conflict of Interest

The authors declare that they have no conflict of interest.

References

[1] H. Oguchi, K. Ishikawa, K. Mizoue, K. Seto and G. Eguchi, Long-term histological evaluation of hydroxyapatite ceramics in humans. *Biomaterials* 1995; 16(1), 33–38.

[2] D. C. Tancred, B. A. O. McCormack and A. J. Carr, A synthetic bone implant macroscopically identical to cancellous bone. *Biomaterials* 1998; 19(24), 2303–2311.

[3] H. Yuan, K. Kurashina, J. D. de Bruijn, Y. Li, K. de Groot and X. Zhang, A preliminary study on osteoinduction of two kinds of calcium phosphate ceramics. *Biomaterials* 1999; 20(19), 1799–1806.

[4] L. Cerroni, R. Filocamo, M. Fabbri, C. Piconi, S. Caropreso and S. G. Condo, Growth of osteoblast-like cells on porous hydroxyapatite ceramics: an in vitro study. *Biomol. Eng.* 2002; 19(2–6), 119–124.

[5] D. W. Hutmacher, J. T. Schantz, C. X. F. Lam, K. C. Tan and T. C. Lim, State of the art and future directions of scaffold-based bone engineering from a biomaterials perspective. *J. Tissue. Eng. Regen. Med.* 2007; 1(4), 245–260.

[6] W. J. E. M. Habraken, J. G. C. Wolke and J. A. Jansen, Ceramic composites as matrices and Scaffolds for drug delivery in tissue engineering. *Adv. Drug. Deli. Rev.* 2007; 59(4–5), 234–248.

[7] M. Taniuichi, H. Takeyema, I. Mizunna, N. Shinagawa, J. Yura, N. Yoshikawa and H. Aoki, The clinical application of intravenous catheter with percutaneous device made of sintered hydroxyapatite. *Jpn. J. Artif. Organs.* 1991; 20, 460–464.

[8] R. V. Silva, J. A. Camilli, C. A. Bertran and N. H. Moreira, The use of hydroxyapatite and autogenous cancellous bone grafts to repair bone defects in rats. *Inter. J. Oral & Maxillofacial Surg.* 2005; 34, 178–184.

[9] A. Stoch, W. Jastrzebski, E. Dlugon, W. Lejda, B. Trybalska, G. J. Stoch and A. Adamczyk, Sol-gel derived hydroxyapatite coatings on titanium and its alloy Ti6Al4V. *J. Mol. Struct.* 2005; 744, 633–640.

[10] S. Best and W. Bonfield, Processing behaviour of hydroxyapatite powders with contrasting morphology. *J. Mater. Sci.: Mater. Med.* 1994; 5(8), 516–521.

[11] K. C. B. Yeong, J. Wang and S. C. Ng, Fabricating densified hydroxyapatite ceramics from a precipitated precursor. *Mater. Lett.* 1999; 38, 208–213.

[12] S. I. Stupp and G. W. Ciegler, Organoapatites: materials for artificial bone. I. Synthesis and microstructure. *J. Biomed. Mater. Res.* 1992; 26(2), 169–183.

[13] T. J. Webster, C. Ergun, R. H. Doremus, R. W. Siegel and R. Bizios, Enhanced osteoclast-like cell functions on nanophase ceramics. *Biomaterials* 2001; 22(11), 1327–1333.

[14] T. J. Webster, R. W. Siegel and R. Bizios, Osteoblast adhesion on nanophase ceramics. *Biomaterials* 1999; 20(13), 1221–1227.

[15] J. K. Thomas, H. Padma Kumar, S. Solomon, C. N. George, K. Joy and J. Koshy, Nanoparticles of $SmBa_2HfO_{5.5}$ through a single step auto-igniting combustion technique and their characterization. *Phys. Status. Solidi A.* 2007; 204(9), 3102–3107.

[16] S. N. Danilchenko, O. G. Kukharenko, C. Moseke, I. Y. Protsenko, L. F. Sukhodub and B. S. Cleff, Determination of the bone mineral crystallite size and lattice strain from diffraction line broadening. *Cryst. Res. Technol.* 2002; 37(11), 1234–1240.

[17] H. P. Klug and L. E. Alexander, *X-ray Diffraction Procedures for and Amorphous Materials.* Wiley, New York, 1974.

[18] Y. J. Wang, J. D. Chen, K. Wei, S. H. Zhang and X. D. Wang, Surfactant-assisted synthesis of hydroxyapatite particles. *Mater. Lett.* 2006; 60(27), 3227–3231.

[19] Y. Wang, S. Zhang, K. Wei, N. Zhao, J. Chen and X. Wang, Hydrothermal synthesis of hydroxyapatite nano-powders using cationic surfactant as a template. *Mater. Lett.* 2006; 60(12), 1484–1487.

[20] R. N. Panda, M. F. Hsieh, R. J. Chung and T. S. Chin, FTIR, XRD, SEM and solid state NMR investigations of carbonate-containing hydrox-yapatite nanoparticles synthesised by hydroxidegel technique. *J. Phys. Chem. Solids.* 2003; 64, 193–199.

[21] J. Xie, X. Meng, Z. Zhou, P. L. Yao, B. X. Gao and Y. Wei, Preparation of titania/hydroxyapatite composite photocatalyst with mosaic structure for degradation of pentachlorophenol. *Mater. Lett.* 2013; 110, 57–60.

[22] T. Giannakopoulou, N. Todorova, G. Romanos, T. Vaimakis, R. Dillert and D. Bahnemann, et al., Composite hydroxyapatite/TiO_2 materials for photocatalytic oxidation of NO_X. *Mater. Sci. Eng. B.* 2012; 177, 1046–1052.

[23] A. Farzadi, M. Solati-Hashjin, F. Bakhshi and A. Aminian, Synthesis and characterization of Hydroxyapatite/b-tricalcium phosphate nanocomposites using microwave irradiation. *Ceram. Int.* 2011; 37, 65–71.

[24] G. E. J. Poinern, R. K. Brundavanam, X. T. Le, P. K. Nicholls, M. A. Cake and D. Fawcett, The synthesis, characterisation and invivo study of a bioceramic for potential tissue regeneration applications. *Sci. Rep.* 2014; 4, 1–9.

Index

About the Authors

Ajay Vasudeo Rane, is currently a doctoral research scholar at Composite Research Group, Durban University of Technology, South Africa. His interest has been in reading and studying the molecular interactions within polymer composites with respect to its mechanical, structural and morphological characterizations. Recycling also has been his topic of interest in near past. He has coedited books with Professor Sabu Thomas, Professor Krishnan Kanny ... specific to recycling of condensation polymers; microscopy applied to life sciences and material sciences.

Amarkumar Bhatt is an eminent chemical engineer working with the research department of Gujarat State Fertilizers and Chemicals. He has a long experience of handling chemical manufacturing plants for fertilizers and industrial products. He had been working with Industrial Waste Incineration Unit of GSFC as well. His expertise is mainly on scaling up processes developed in the laboratory scale to pilot scale.

Arpita Das, has completed her Master's degree in Food and Nutrition from University of Calcutta in 2010 and is working as teaching faculty in a college affiliated to Burdwan University, WB, India since 2016. She just have started her research carrier with exploration of N, P content in crops of waste land compared to normal agricultural land. She also have interest in Nutritional genomics and have started this part of work also.

Bangari Daruka Prasad is working on synthesis of energy saving, ecofriendly, efficient single and multi-phased rare earth doped ions for display devices and radiation monitoring. He has published about 87 research papers in peer reviewed and high impact journals includes *Crystal Growth & Design, ACS Sustainable Chemistry and Engineering, Alloys and Compounds,* etc. His research contributions have earned him 1721 scientific citations with H-index of 25 and I-index of 56. Received the research fund for the year 2016–17 of Rs. 5 lakhs from Vision Group of Science and Technology,

Government of Karnataka, under the scheme research fund for talented teachers (RFTT). He is a life member of Materials Research Society of India, ISTE, Solar Energy Society of India, Luminescence Society of India etc. He has published five books on Engineering Physics published in Acme-Learning. He is also a guest editor for Journal of Nanomaterials, Hindwai Publications.

Christopher Thresiamm Mathew is an Associate Professor of Physics, Mar Ivanios College, University of Kerala, Trivandrum, India. He pursued his PhD Physics degree from University of Kerala in 2017. His expertise lies in the area of infrared transparent ceramic materials and nanomaterials. He has published more than 10 research papers in various journals and conference proceedings.

Dhamodaran Dhanalakshmi had worked as scientific assistant at Raman Research Institute. She worked in experiments with cold atoms and quantum optics and has interest in science teaching.

Dibyendu Roy is a theoretical physicist with research interests in condensed matter, non-equilibrium statistical mechanics and atomic, molecular and optical (AMO) physics. Currently he is an associate professor at Raman Research Institute.

Elias Chatzitheodoridis is a Professor in Mineralogy and Petrology at the National Technical University of Athens (NTUA), Greece. He acquired his first degree in Geology in 1987 from the National and Kapodistrian University of Athens, Greece. He acquired his MSc in 1990 and PhD in 1994 both from the University of Manchester, Department of Earth and Environmental Sciences. A change of his career path was made towards new technologies, such as micro- and nanofabrication, micro-assembly of microelectromechanical systems (MEMS), lithography, and photonics. He worked in the telecom industry. Returning back to Earth and Planetary sciences, he developed further his original research interests in the area of alteration minerals and environments found in Martian meteorites, origins of life studies and astrobiology, detection of biosignatures using state-of-the-art instrumentation and analytical methods.

Hanumanthappa NagBhushana M.Sc. Physics–1997, Bangalore University, Bangalore, Karnataka, India. M.Phil. Physics–2000, Bangalore

University. Ph.D. Solid state Physics–2002, Bangalore University, D.Sc. in Materials Science, Tumkur University, Tumkur, Karnataka, India, Bangalore University. Presently working as Professor and Head of the Department, Prof. CNR Rao Centre for Advanced Materials, Tumkur University, Tumkur-572103, India. Area of Interest: Nanomaterials, Luminescence, defects studies, SHI irradiation, Space Science, etc. Till date: Citations-6395, H-index-41 and I-index-183. Reviewer for many journals and published many number of book/book chapters through reputed Publishers. He completed many funded projects funded by many government and private agencies. Under his guidance 14 students awarded for their Ph.D. degree. He is the potential reviewer for many international, funding agencies, journals and books. He is the life member for many international reputed societies.

Hema Ramachandran is an experimental physicist and currently a professor at Raman Research Institute. Her research interests include light in random media, single atom traps, photon statistics and brain computer interfaces.

Jayachandran Santhakumari Lakshmi is a Ph.D research scholar in Materials Science and Nanotechnology at the University of Kerala, India. She pursued a Master's degree in Physics and qualified The National Eligibility Test (NET). Currently she is concentrating in low temperature co-fired ceramic at the Electronic Materials Research Laboratory, Department of Physics, Mar Ivanios College (Autonomous), University of Kerala.

Jesiya Susan George is a Research Scholar at International and Inter-University Centre for Nanoscience and Nanotechnology (IIUCNN), Mahatma Gandhi University Kottayam, India. She completed Master of Science in Bio-polymer Science from Cochin University of Science and Technology, Kochi, Kerala, her area of research involves Epoxy/Graphene Oxide composites. She wrote a chapter on **"Manufacturing of slow and controlled release pesticides"** published by **Springer International Publishing.**

Jijimon K. Thomas is Associate Professor of Physics, Mar Ivanios College, University of Kerala, Trivandrum, India. He has presented papers in many national and international seminars and also published more than 100 international papers and many patents in his credits. He pursued his PhD Physics degree from CSIR Regional Research Laboratory – Trivandrum in 1997 Physics and M Inst P from Institute of Physics, London in 2010. His expertise lies in the area of Electronic Ceramic materials, nano-materials and

Superconductivity. In 2001 he have initiated the establishment of a research wing "Electronics Materials Research Laboratory (EMRL)" associated with the department of physics, Mar Ivanios College to promote materials science research which correlates the materials science research in the college. The laboratory has procured many major instrumental set up and presently 10 PhD students are doing research. His group has published more than 104 research papers in SCI journals and about 52 conference proceedings. In 2000 his group possessed three US patents, one European patent and 2 Indian patents, for developing new ceramic substrates for superconductors and group had reported the XRD patterns of 10 compounds in JCPDS file. He is concentrating on the development of perovskites and scheelite group of materials and IR transparent ceramics in nanostructured form to tailor their dielectric, optical and structural properties in order to suit their applications as electronic, dielectric and optoelectronic materials. Apart from these, study on the improved critical current density of superconductors by incorporating nanostructured artificial flux pinning centers is also an area of his research.

Keloth Paduvilan Jibin Currently working as JRF under DST Nanomission Project at IIUCNN, and PhD Fellow at School of Chemical Sciences, Mahatma Gandhi University, Kottayam. Qualified UGC CSIR NET in 2016 (National level exam). Secured first rank in MSc Analytical Chemistry 2015–2016, Mahatma Gandhi University, Kottayam, Kerala. Participated and paper presented in international conferences ICMS 2017, ICECM 2017, ICRM 2018, ICN 2018, ICMST 2018, ICRAMC 2019, ICSG 2019 etc. Participated in several national seminars on different areas of chemistry. It includes a workshop for chemistry students and teachers on September 2014 jointly coordinated by JNCASR and NIIST at Trivandrum. Worked on a project entitled *"optical studies of samarium complex synthesised using curcumin isolated from turmeric extract as ligand* and also in the project *synthesis of Mg and Co codoped ZnO based diluted magnetic semi-conductor for optoelectronic applications.* ***Edited one book titled as Reuse and Recycling of Materials published by River Publishers.***

Khateef Riazunnisa, working as Assistant Professor, Dept of Biotechnology and Bioinformatics, Yogi Vemana University, Kadapa, pursued Ph.D. *Plant Sciences*, from University of Hyderabad, Hyderabad. She received DST Young Scientist award in 2013. She visited Germany to carry out research work at University of Osnabrueck in 2006. She has qualified

GATE and UGC-CSIR – JRF-SRF, eligibility for lectureship in June 2003. Dr. Riazunnisa published 24 research papers in Peer Reviewed International and national journals and presented 40 papers in International and national conferences/seminars. Four seminars were organized by her. She has completed two research projects from DST-SERB and Agri Sci Park. Under her supervision one student was awarded and two students are pursuing Ph.D. She is the Journal Reviewer for Natural Product Research (Taylor and Francis), African Journal of Biotechnology, African Journal of Pure and Applied Chemistry, Industrial crops and products (Elsevier publishers).

Krishnan Kanny, is a full Professor at Department of Mechanical Engineering, Durban University of Technology, South Africa. Krishnan Kanny is a seasoned engineer with over 34 years of experience in leadership, management and human resource development. He received his Ph.D. in Material Science and Engineering from Tuskegee University, Alabama, United States. He is an NRF rated scientist. Krishnan Kanny's research interest includes designing, processing and testing of composites for structural and functional applications. Computational modelling and analytical modelling are his other areas of research and interest. Krishnan Kanny has five edited books and more than 150 research articles to his credit.

Krishnan Thyagarajan, working as Professor and Head at JNTUA College of Engineering, Pulivendula. He is having more than 25 years of teaching and research experience. To his credit there are 55 international journal publications, 21 international conference and 30 national conference proceedings. Six books are published to his credit and nine students received their Ph.D degree under his guidance. He is a Chairman, member for many academic boards of JNTUA and other universities. He organized four workshops and many invited lectures delivered in national and international level. He is the life member of many scientific societies.

Maheswar Swar is pursuing his Ph.D in experimental physics at Raman Research Institute. His research interests include quantum optics, atomic, molecular and optical (AMO) physics, precision measurements, Laser cooling and trapping with a special focus in studying spin properties of such systems.

Monalisa Mishra is Assistant professor of Department of Life Science in National Institute of Technology Rourkela, India. She has earlier served as

Assistant Professor in the Department of Biological sciences, BITS, Pilani (2012–2014). Her research interest is in developmental biology and neuroscience with specialization in structural analysis of eye of various arthropod families, ultra structure analysis of vertebrate eye (both mice and Zebra fish) and expertise in *Drosophila* eye morphogenesis. She is specialized in electron microscopy and confocal microscopy. She has more than 7 years of teaching experience in General biology, Animal physiology, Bio lab, Instrumental method analysis, Developmental biology.

She has also published more than 60 research articles in the peer-reviewed international journal and authored or co-authored numerous books and book chapters.

Neehara Alackal, completed her master's degree in chemistry at Department of Chemistry, Dev Matha College, Kuravilangad affiliated to Mahatma Gandhi University, Kottayam. Topics that catches her interest of reading includes organic chemistry, analytical techniques for characterizations.

Nirmit Kantilal Sanchapara holds his bachelor's degree in Mechanical Engineering from CHARUSAT University, India. He has been working as a Research Engineer with Gujarat States Fertilizers and Chemicals. Fertilizer process design and Techniques of Granulation are his key area of expertise. He holds an Indian patent for Waste Heat Air-conditioning System. He was the lead member of the team which bagged the prestigious Go Green Award of SAE SUPRA 2014 competition held in India. He has expertise in development of software like "Parametric Modelling" and "Development of Customised Fertilizers".

Prajitha Velayudhan is currently working as Research Scholar at School of Chemical Sciences and JRF under DST Nanomission Project at International and Inter University Centre for Nanoscience and Nanotechnology, Mahatma Gandhi University, Kerala, India. She completed Master of Science in Biopolymer Science from Cochin University of Science and Technology, Kochi, Kerala. Her research focuses on self-assembling of nanomaterials such as nanosilica, Graphene oxide and Rubber Nanocomposites etc. *Edited one book titled as Colloidal Metal Oxide Nanoparticles published by Elsevier Publishers.*

Sam Solomon is currently working as an associate professor in Mar Ivanious college, Thiruvananthapuram. He has presented papers in many national and international seminars and also published more than 100 international papers and many patents in his credits.

Sandeep Jasvantrai Parikh currently works as a Senior Vice President with Gujarat State Fertilizers and Chemicals, India. After obtaining his Bachelor's Degree in Chemical Engineering from Gujarat University, India, he joined GSFC in 1984 to work with various fertilizer and industrial product manufacturing plants.

Sandeep Kollam is a Research Scholar in the Department of Physics, Mar Ivanios College, Thiruvananthapuram. He has published many international papers in the field of nanoceramics, and also presented papers in many national and international seminars and workshops.

Sanjukta Roy is an experimental physicist working at Raman Research Institute. Her research interests include mesoscopic quantum entanglement, quantum statistics, Bose-Einstein condensation, few body physics and ultra-cold atoms.

Saptarishi Chaudhuri is an experimentalist and currently an associate professor at Raman Research Institute. His research interests are Quantum degenerate atomic and molecular systems, quantum statistics, Laser cooling and trapping of atoms, precision measurements, Quantum magnetism and Quantum simulation of complex systems.

Sohan Jheeta, FRSB, FRMS, FRAS is a UK based independent educator, science communicator and research scientist. Gained his PhD later in life, in 2010: it was in physics, specifically in astrochemistry as applied to the origin of life and the thesis was entitled: "Experimental Studies in Prebiotic Chemistry" under the guidance of Professor Nigel Mason OBE of the Open University, UK. He runs two scientific groups, namely the Network of Researchers on the Chemical Evolution of Life (NoR CEL) and the Frontiers of Sciences—the latter group is involved in disseminating "science-in-society" specifically to the public. For full details of his work please visit the website www.sohan@sohanjheeta.com.

Soheb Husenmiyan Shekh has vast experience in chemical synthesis research and analysis. After completing his master's degree in Organic Chemistry as a University Topper from Gujarat University, India he has been working at various laboratories as a synthesis chemist and analyst. His areas of interest in research are spectroscopy, organic synthesis, catalysis etc.

Steffy Maria Jose is a Ph.D research scholar in Materials Science and Nanotechnology at the University of Kerala, India. Prior to her present affiliation at the University of Kerala, she earned a Master's degree and M.Phil degree in Physics both from Mahatma Gandhi University, Kerala, India. She gained research interest during her Master's and M.Phil programme in Physics and gained basic knowledge in research methodology. Currently she is doing research in the synthesis of yttria based nanoceramics and composites for their applications as infrared transparent window materials at the Electronic Materials Research Laboratory, Department of Physics, Mar Ivanios College (Autonomous), University of Kerala.

Suresh Puthiyaveetil Othayoth obtained his PhD from the Department of Environmental Sciences of Macquarie University in Sydney, Australia. He did his post doctoral research at Macquarie University and then at Academia Sinica in Taiwan. Currently he works with the R&D Centre of Gujarat State Fertilizers and Chemicals Limited in India. His research interests include soil evolution, plant nutrients, earth surface processes, analytical techniques etc. He has published many research papers in peer reviewed journals and presented research papers in reputed international conferences. His current research focuses on development of method for supplying soil nutrients effectively for soil quality enhancement.

Syeda Anjum Mobeen and **Maram Vidya Vani** are research scholars under the supervision of Dr. K. Riazunnisa, Assistant Professor, Department of Biotechnology and Bioinformatics, Yogi Vemana University, Kadapa, India. SAM is recipient of UGC-MANF fellowship. Her research area is Nanotechnology, Bio-informatics (drug design). She published two articles in International and one National journals, two Book chapters. She also presented 20 papers in International, National, workshops/seminars/conferences. MVV is working on Bio-fuel production, development of light efficient algal mutants. She published two papers and also presented 9 papers in International, National, workshops/seminars/conferences.

Tandrima Chaudhuri, working as teaching faculty in a college affiliated to Burdwan University WB, India since 2005, has 16 years research experience in various field of Chemistry viz. Material Science,

Photochemistry, Computational Chemistry. She has over 40 international publication cited regularly. Sensing of cation and anion along with DNA compaction-decompaction studies are new interest of field to explore in research in her credit.

V. K. B. Kota, born in March 1951 in Kakinada, A.P. (India), did his Ph.D. in physics from Andhra University, Visakapatnam (India) in 1977 and became a faculty member in the Theoretical Physics Division of Physical Research Laboratory (PRL), Ahmedabad in 1980. He has superannuated from PRL in 2013 and continuing there as a Honorary faculty member. He was Research Associate at Department of Physics and Astronomy, University of Rochester, Rochester (N.Y.), U.S.A. during 1980–83 and Senior Research Associate during 1987–88, visiting scientist at Laurentian University, Sudbury, Canada during January–March, 1996 and at Max-Planck Institute for Nuclear Physics, Heidelberg, Germany during May–June 1999. He held the position of adjunct Professor in the Department of Physics, Laurentian University, Sudbury, Ontario, Canada during 2007–2015.

Prof. Kota has guided 3 Ph.D. students and had more than 12 post doctoral fellows. He has published more than 150 papers in international and national journals and also published eight review articles with two of them in Physics Reports and one in Annual Reviews of Nuclear and Particle Science. He has published laboratory reports with more than 850 pages. In addition, he has authored three books: (i) Statistical Spectrocopy with R.U. Haq (World Scientific); (ii) Embedded Random Matrix Ensembles (Springer); (iii) Structure of Medium Mass Nuclei with R. Sahu (CRC press of Taylor & Francis). He has given invited talks in large number of international and national conferences/symposiums/schools/workshops. He has received Indian National Science Academy Gold Medal for Young Scientists for the year 1981. Dr. Kota was a Member: (i) Programme Advisory Committee (PAC), Department of Science and Technology (Government of India) on Plasma, High Energy, Nuclear Physics, Astromomy and Astrophysics and Non-linear Dynamics for the period 2012–2015; (ii) Member, Board of Studies in Physics and Meteorology and also member of Faculty Board of M.S. University of Baroda, Vadodara, India for the period 2011–2014; (iii) Member, Editorial Board for the journal "Journal of Nuclear Physics, Material Sciences, Radiation and Applications (JNPMSRA)" from its inspection in 2013 [Journal Website: http://jnp.chitkara.edu.in]. Also, since October 2001, Reviewer for 'Mathematical Reviews' database of American Mathematical Society.

Yesoda Velukutty Swapna is a Ph.D research scholar in Materials Science and Nanotechnology at the University of Kerala, India. She pursued a Master's degree and M.Phil degree in Physics before joining PhD. Currently she is doing research in the ceramics for their applications as scaffolds materials at the Electronic Materials Research Laboratory, Department of Physics, Mar Ivanios College (Autonomous), University of Kerala.

About the Editors

Yang Weiman is Professor of Mechanical and Electrical Engineering at Beijing University of Chemical Technology, and directing the Polymer Processing and Advanced Manufacturing Center. He is also a Distinguished Professor of Chang Jiang Scholars Program, China Ministry of Education. His research interest is polymer processing and advanced manufacture technology, mainly focused on green process of polymer producing, plastics precision molding, energy saving in tire manufacturing, nano-fiber electrospinning and enhanced heat transfer in polymer processing, etc. His research group has undertaken over 30 projects supported by National Science and Technology Support Plan, National Science Foundation of China, and industrial company. Based on the research results, he has applied more than 200 invention patents (82 items authorized to present), published 8 books (English: Advances in Polymer Processing, published by Woodhead in UK and CRC in USA) and more than 300 journal papers. He has received numerous honors and recognitions, including 2 Awards of China National Science and Technology Progress, 10 China Provincial Awards and Hou te Pang Chemical Science and Technology Award. Dr. Yang holds a B.S. degree of Mechanical Engineering, and a M.S. and a Ph.D. Degree of Chemical Process Equipment from Beijing University of Chemical Technology in 1987 and 1990, and 1998, respectively. From Oct. 2000 to Sept. 2002, he was a postdoctoral fellow at Polymer Processing Yokoi Lab. of the University of Tokyo, Japan. In the recent years, Dr. Yang has been serving on editorial board of some important journals related to polymer processing, and also selected to Vice Chairman of the Experts Committee of China Plastics Processing Industry Association, and Advisor Experts of China Rubber Processing Industry Association.

Jibin K. P. Currently working as JRF under DST Nanomission Project at IIUCNN, and PhD Fellow at School of Chemical Sciences, Mahatma Gandhi University, Kottayam. Qualified UGC CSIR NET in 2016 (National level exam). Secured first rank in MSc Analytical Chemistry 2015–2016, Mahatma

Gandhi University, Kottayam, Kerala. Participated and paper presented in international conferences ICMS 2017, ICECM 2017, ICRM 2018, ICN 2018, ICMST 2018, ICRAMC 2019, ICSG 2019 etc. Participated in several national seminars on different areas of chemistry. It includes a workshop for chemistry students and teachers on September 2014 jointly coordinated by JNCASR and NIIST at Trivandrum. Worked on a project entitled *"optical studies of samarium complex synthesised using curcumin isolated from turmeric extract as ligand* and also in the project *synthesis of Mg and Co codoped ZnO based diluted magnetic semi-conductor for optoelectronic applications.* **Edited one book titled as Reuse and Recycling of Materials published by River Publishers.**

G. L. Praveen received his Ph.D in Chemistry under the guidance of Dr. Sony George, Head, Department of Chemistry, University of Kerala, India in the year 2016. He cleared Master of Philosophy degree in the year 2010 from the same department. He was involved in the Synthesis and characterization of various nanostructures which find profound application in the field of cancer treatment and sensor materials. In the year 2017, he qualified the Post-Doctoral Research (DST SERB NPDF) fellowship under the mentorship of Professor Dr. Sabu Thomas, Mahatma Gandhi University, India. During his research venture, He worked as a project fellow in CSIR-NIIST (National Institute for Interdisciplinary Science and Technology) Thiruvananthapuram, India, Sree Chitra Tirunal Institute for Medical Sciences and Technology, Thiruvananthapuram, India and Mar Ivanios College, Thiruvananthapuram. At present he joined to Wimpey Laboratory, Dubai, as senior analytical Chemist.

Professor Sabu Thomas is currently the **Vice Chancellor of Mahatma Gandhi University** and the Founder Director and Professor of the International and Interuniversity Centre for Nanoscience and Nanotechnology. He is also a full professor of Polymer Science and Engineering at the School of Chemical Sciences of Mahatma Gandhi University, Kottayam, Kerala, India. Prof. Thomas is an outstanding leader with sustained international acclaims for his work in Nanoscience, Polymer Science and Engineering, Polymer Nanocomposites, Elastomers, Polymer Blends, Interpenetrating Polymer Networks, Polymer Membranes, Green Composites and Nanocomposites, Nanomedicine and Green Nanotechnology. Dr. Thomas's ground breaking inventions in polymer nanocomposites, polymer blends, green bio-nanotechnological and nano-biomedical sciences, have made transformative differences in the development of new materials for automotive, space, housing and biomedical fields. In collaboration with India's premier tyre company, Apollo Tyres, Professor Thomas's group invented new high performance barrier rubber nanocomposite membranes for inner tubes and inner liners for tyres. Professor Thomas has received a number of national and international awards which include:

Fellowship of the Royal Society of Chemistry, London FRSC, Distinguished Professorship from Josef Stefan Institute, Slovenia, MRSI medal, Nano Tech Medal, CRSI medal, Distinguished Faculty Award, Dr. APJ Abdul Kalam Award *for* ***Scientific Excellence – 2016, Mahatma Gandhi University – Award for Outstanding Contribution*** *– Nov. 2016,* ***Lifetime Achievement Award of the Malaysian Polymer Group, Indian Nano Biologists award 2017*** and ***Sukumar Maithy Award*** for the best polymer researcher in the country. He is in the list of **most productive researchers in India** and holds a position of No. 5. Because of the outstanding contributions to the field of Nanoscience and Polymer Science and Engineering, Prof. Thomas has been conferred ***Honoris Causa (D.Sc) Doctorate by the University of South Brittany***, Lorient, France and ***University of Lorraine, Nancy, France.*** Very recently, Prof. Thomas has been awarded ***Senior Fulbright Fellowship*** to visit 20 Universities in the US and most productive faculty award in the domain Materials Sciences. Very recently he was also awarded with ***National Education Leadership Award – 2017*** for Excellence in Education. Prof Thomas also won **6th** ***contest of "mega-grants"*** in the grant competition of the Government of the Russian Federation ***(Ministry of Education and Science of the Russian Federation)*** designed to support research projects implemented under the supervision of the world's leading scientists. He has been honoured with ***Faculty Research Award*** of India's brightest minds in the field of academic research in May 2018. Professor Thomas was awarded with ***Trila – Academician*** of The Year in June 2018 acknowledging his contribution to tyre industry. This year, Prof Thomas was also awarded with ***H.G. Puthenkavu Mar Philoxenos Memorial Best Scientist Award.*** In 2019 Prof. Thomas received prestigious ***CNR Rao prize lecture Award.*** Prof. Thomas has published over 900 peer reviewed research papers, reviews and book chapters. He has co-edited 117 books published by Royal Society, Wiley, Wood head, Elsevier, CRC Press, Springer, and Nova etc. He is the inventor of **5 patents.** The **H index** of Prof. Thomas is **97** and has more than **43,871 citations.** Prof. Thomas has delivered over 300 Plenary/Inaugural and Invited lectures in national/international meetings over 30 countries.

Nandakumar Kalarikkal is a Professor, Chair and Director at School of Pure and Applied Physics of Mahatma Gandhi University, Kottayam, Kerala, India. He is also the Director of International and Inter University Centre for Nanoscience and Nanotechnology of Mahatma Gandhi University, Kottayam, Kerala, India. His research activities involve applications of nanostructured materials, laser plasma, phase transitions, etc. He is the recipient of research fellowships and associate ships from prestigious government organizations such as the Department of Science and Technology and Council of Scientific and Industrial Research of Government of India. He has active collaboration with national and international scientific institutions in India, South Africa, Slovenia, Canada, France, Germany, Malaysia, Australia and US. He has more than 160 publications in peer reviewed journals. He has also co-edited 15 books of scientific interest and co-authored many book chapters.